66th and 67th Annual Meeting of the Labor and Employment Relations Association, LERA@ASSA Meeting 2014 and LERA@ASSA Meeting 2015

Labor and Employment Relations Association Series

Philadelphia, Pennsylvania, USA
3-5 January 2014

Portland, Oregon, USA
29 May-1 June 2014

Boston, Massachusetts, USA
3-5 January 2015

Pittsburgh, Pennsylvania, USA
28-31 May 2015

Editors:

Ariel Avgar

ISBN: 978-1-7138-0275-4

Printed from e-media with permission by:

Curran Associates, Inc.
57 Morehouse Lane
Red Hook, NY 12571

Some format issues inherent in the e-media version may also appear in this print version.

Labor and Employment Relations Association (LERA)
University of Illinois at Urbana-Champaign
121 Labor and Employment Relations Bldg.,
504 E. Armory Ave.
Champaign IL 61820
USA

Phone: (217) 333-0072
Fax: (217) 265-5130

LERAoffice@illinois.edu

Additional copies of this publication are available from:

Curran Associates, Inc.
57 Morehouse Lane
Red Hook, NY 12571 USA
Phone: 845-758-0400
Fax: 845-758-2633
Email: curran@proceedings.com
Web: www.proceedings.com

LABOR AND EMPLOYMENT RELATIONS

ASSOCIATION SERIES

Proceedings of the LERA 2014–2015 Meetings

LERA@ASSA Meeting
January 3–5, 2014, Philadelphia, Pennsylvania
(in conjunction with ASSA/AEA)

LERA 66th Annual Meeting
May 29–June 1, 2014
Portland, Oregon

LERA@ASSA Meeting
January 3–5, 2015, Boston, MA
(in conjunction with ASSA/AEA)

LERA 67th Annual Meeting
May 28–31, 2015, Pittsburgh, PA

Ariel Avgar, Editor-in-Chief

LABOR AND EMPLOYMENT RELATIONS ASSOCIATION SERIES
(Formerly the Industrial Relations Research Association)

Proceedings of the X nnual Meeting
Annual Research Volume
Perspectives on Work Magazine

Inquiries and other communications regarding membership, meetings, publications, and general affairs of the Association, as well as notice of address changes, should be addressed to the LERA office.

LABOR AND EMPLOYMENT RELATIONS ASSOCIATION
University of Illinois at Urbana-Champaign School of Labor and Employment Relations
121 Labor and Employment Relations Building, MC-504
504 East Armory Avenue, Champaign, IL 61820
Telephone: 217/333-0072 Fax: 217/265-5130 E-mail: LERAoffice@illinois.edu
Website: http://lera.memberclicks.net EPRN: http://www.employmentpolicy.org/

CONTENTS

VI.
LERA PAPERS I: NEW MODELS

Eric Duchinsky, Presiding

VII.
LERA PAPERS VI: LONGEVITY

Eric Duchinsky, Presiding

VIII.
LERA ANNUAL REPORTS

LABOR AND EMPLOYMENT RELATIONS ASSOCIATION SERIES

Proceedings of the LERA 2014–2015 Meetings

LERA@ASSA Meeting
January 3–5, 2014
Philadelphia, Pennsylvania

LERA 66th Annual Meeting
May 29–June 1, 2014
Portland, Oregon

LERA@ASSA Meeting
January 3–5, 2015
Boston, Massachusetts

LERA 67th Annual Meeting
May 28–31, 2015
Pittsburgh, Pennsylvania

2014 Presidential Address

Renewing Employment Relations for Shared Prosperity

LISA M. LYNCH
Brandeis University

"We are back to where many of us were 20 years ago—still grappling with the problem of how to simultaneously reduce unemployment, avoid inflation, and try to maintain a free society … We need to develop selective employment policies in order to be able to deal with it".[1]

With these words, U.S. Secretary of Labor Ray Marshall finished his Presidential address to the Industrial Relations Research Association, what we now call the Labor and Employment Relations Association. The year was 1977 and I was at that address—hiding behind a potted plant in the back of a huge ballroom in New York City. I was a senior at Wellesley College attending the meeting with my then boyfriend now husband, Fabio Schiantarelli. We didn't have the courage or the money to sit at any table—too many famous government officials, professors, union Presidents, and business leaders. Thus began my 37 year old relationship with this wonderful organization.

How did I end up at that meeting? It all begins with what I witnessed as a teenager growing up in the early 1970s in Waterbury, Connecticut and then Buffalo, New York. My father was a metallurgist who worked as a research scientist and then manager at a number of brass and copper mills beginning in the 1950s. As a kid I loved to visit the mill where he worked—there is nothing more exciting than seeing big furnaces and hot metal being rolled. The noise, the smell of burning oil and metal, the dirt—it was great. However, throughout the 1970s I watched as one manufacturing plant after another shut down in Waterbury and Buffalo. I saw displacement, unemployment and hardworking people stymied by their lack of skills or paucity of alternative jobs. Watching my father and the people he worked with deal with these challenges, ultimately helped shape my career choice.

When I got to college I wanted to find a way to prepare myself for a professional life where I would be able to re-shape the options for workers facing unemployment. Fortunately I found a mentor in my first year at Wellesley—Phyllis Wallace, a professor of industrial relations at MIT's Sloan School of Management, a former IRRA President, and the first woman professor at Sloan (for more information on her extraordinary life read Malveaux (1994)). Personally, she had broken through dual discrimination barriers as an African American and woman and contributed path breaking work as a scholar and practitioner to open up the labor market to and minorities. She was the one who told me to go and listen to Ray Marshall speak. She said he was going to be one of the great Secretaries of Labor. She was right—she was always right.

In comparing and contrasting the nature of labor and employment relations at the end of the 1970s and early 1980s to today, there are important difference and similarities. Unemployment in December 1977 was 6.4 percent; remarkably similar to our current rate of unemployment 6.3 percent.

Unemployment rates in the late 1970s were highest for teens, blacks and Hispanics, just as they are today. And those with less than a high school degree had more difficulty finding work than those with a college degree—again, just like today.

But there are important differences between 1977 and now. The first is the changing gender composition of the workforce. In 1977, 78 percent of working age men were in the labor force. Today just 69 percent are. Over the past 35 years as men delayed entry into work or retired earlier, women instead increased their labor force participation. In 1977, 49 percent of working age women were in the labor force while today 57 percent are working. This change is due to a wide range of factors starting with education. In 1977 a higher percentage of men compared to women 18-24 years of age were enrolled in college. Today, 72 percent of high school graduates who are women enroll in college versus 65 percent of men. Over the past decade, more women than men in the labor force (aged 25-64) have a college degree. With this explosion of educational attainment we have seen a surge in the percentage of women in the labor force (see Goldin (2006), and Goldin, Katz, and Kuziemko (2006) for more discussion on this).

Additional factors have also influenced the rising participation of women. These include labor saving devices that freed women from household chores (see Greenwood, Seshadri and Yorukoglu (2005)); improved maternal health and improved survival of children (see Albanesi and Olivetti (2014)); the pill (see Goldin and Katz (2002)); divorce (Peters (1986)); tax credits and child care (Averett, Peters and Waldman (1997)); and changing attitudes (for example the finding that men whose mother worked were more likely to have working wives in Fernández, Fogli, and Olivetti (2004)). Changing government policies, business practices, and legal decisions over the past 35 years have also been critical in supporting this increased participation. New data and the scholarly work of many members of LERA have made a contribution in terms of documenting these changes and seeing what policies have worked to advance participation in the labor market.

With increased education and more work experience women have seen their wages rise relative to men. In 1977 women who were working full time earned on average 59 percent what men earned. Today they earn 77 percent. This is still not 100 percent and part of this gap can be explained by remaining differences in industry or occupation, work experience, union status, and part time status. But an "unexplained" gap persists and there is ongoing debate over whether or not this is evidence of discrimination, some "cultural" preference, or that workers "pay" to have workplace flexibility. In addition, much of the narrowing of the gap in wages between men and women in recent years has been due to stagnant or falling male wages—not something I would score as a victory for either men or women.

While the 1980s and 1990s were characterized by rising labor force participation of women, as we entered the "Great Recession" we saw the labor force participation rate for both men and women drop—the decline was especially dramatic for men—thus the reference to the so-called "mancession". For men this was attributed to their higher concentration in the construction industry. But labor force participation had leveled off and declined even before the recession for women. In 1990, the US ranked 6th in female labor force participation in advanced industrialized economies. But by 2010 we ranked 17th. What happened? Francine Blau and Lawrence Kahn (2013) present compelling evidence that many other countries expanded social support structures over the last twenty years and this helped more parents remain in the labor market. These social supports included expanded sick leave coverage, state supported high quality child care, expansion of paid parental leave, and protections for workers from unpredictable shift work schedules. Unfortunately, we have chosen not to expand such supports in the United States. In the US married full time working women spend almost 50 percent[2] more time than their spouses doing housework and taking care of children or other aging family members—the so called second shift. Women single heads of household with children face similar challenges to married working women in terms of availability of affordable quality child care and after school programs. But many single women with children also face the challenge of being employed in sectors of our economy that increasingly have more unpredictable shift schedules that can wreak havoc in their ability to have their children well cared for while they are at work.

There are two additional explanations offered to explain this decline in the labor force participation rate: 1.) delayed entry of youth into labor market and (2.) the aging of the workforce. As I mentioned earlier, young people have been particularly hard hit by this recession. Almost one in five teens are unemployed today compared to 15.5 percent in 1977. But more striking is what has happened to the fraction of teens in employment. In 1977 almost 48 percent of teens had a job. Today only 27 percent of teens have a job. More youth are delaying entry into the labor market and are staying on in school. This is a good thing in terms of the return to education. But there are many people leaving college who can only find unpaid internships and even more disturbing a growing fraction of youth who are not in school or the labor force. In 2003, well before the Great Recession took its toll on youth employment, this group of disconnected youth was estimated by Andrew Sum and Ishwar Khatiwada (2004) to be over 5.7 million. These young people represent an enormous waste of human potential. Our failure to connect these young people to school and work will haunt us will into the future.

At the other end of the age distribution our workforce is aging. In the late 1970s the average age of a worker was 35—today it is 42. As baby boomers have "greyed" the share of the labor force over the age of 55 has risen from just over 10 percent in 1990 to almost 20 percent in 2010. By 2020 one in four workers will be over the age of 55. The labor force participation rate has historically been much lower for workers over the age of 55 so as a result the overall labor force participation rate falls as the workforce ages.

How much the overall labor force participation rate will fall due to the aging of the workforce is still a matter of debate. For example, the labor force participation rate for those 65-69 rose 28 percent over the recession and 41 percent for those over 70. Some of this is due to the lack of savings and pensions so that workers who would have liked to have retired could not. But some of this reflects improving health, more education and a desire to continue working. Baby boomers have defied predictions by labor economists about their labor market behavior when they were young and I suspect will continue to surprise us as they redefine what it means to retire. This will also be an area where companies will need to adjust their workplace practices.

Not all of the recent fall in the labor force participation rate can be explained just by demographic trends. There are also cyclical factors in play. As I discussed earlier, the unemployment rate in December 1977 and today are almost identical. But the nature of unemployment has changed profoundly. In 1977 a little over 13 percent of those unemployed had been out of work for 6 months or more. Most workers at that time were on temporary layoff. Today more than 35 percent of unemployed workers have been out of work for more than 6 months. As workers remain in long spells of unemployment they are more likely to end up dropping out of the labor force due to discouragement with finding employment. You are only counted as unemployed in the US if you have actively searched for work over the past four weeks. After exhausting all possible employment possibilities in one's town it is not surprising that workers might stop looking for work. Workers will know when employment comes back.

In principal then, as labor market conditions improve we should see more people returning to the labor market who had dropped out. This would increase the labor force participation rate and even possibly the unemployment rate in the short run.

However, many workers in long spells of unemployment may have been scarred by this experience. My work on youth unemployment on youth labor markets in the late 1970s and 1980s (see Imbens and Lynch (2006) and Lynch (1983, 1985, 1989) suggested that entering a labor market when unemployment was high had a permanent negative effect on the subsequent labor market experience of youth—both in terms of their wages and the number of weeks worked. We also know that long spells of unemployment are associated with worsening physical and mental health so we should not be surprised at the marked increase in applications for permanent disability insurance. Employers may be more reluctant to take someone on who has been out of work a long period of time, workers lose their networks and contacts that might help them find new employment, and workers can become so discouraged they drop out. As a result, some economists are arguing that the current unemployment rate of 6.3 percent is the new "full employment" rate because workers are now permanently out of the labor market.

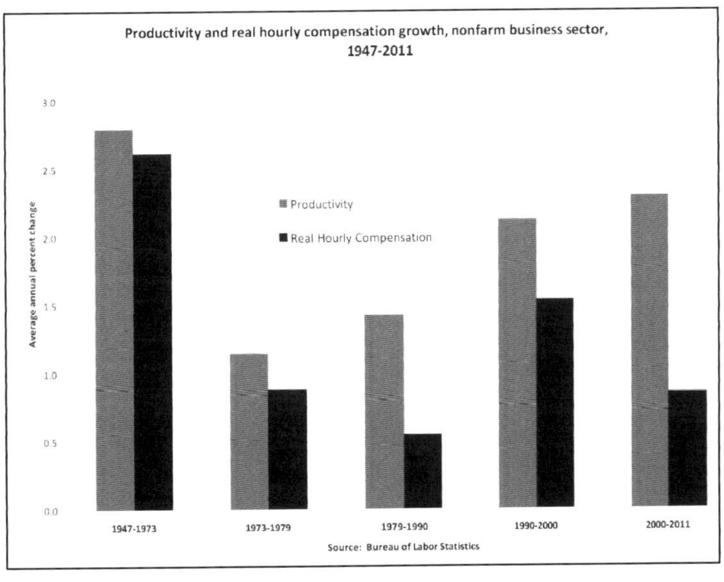

What does this mean in terms of policy responses? If you believe the story that those of working age who have stopped looking for work will never return to the labor force because of discouragement or skills mismatch then there is no need to support investments in government training, unemployment insurance, or expansionary monetary or fiscal policy. The evidence to support this structural story of unemployment is weak. If we had a skills mismatch then we should be seeing increases in wages for workers in highly demanded jobs. But we do not. In the first quarter of 2014 there were approximately 3 unemployed workers for every job vacancy. So what we need is jobs!

This brings me to the final fundamental change in the labor market of today compared to that of the late 1970s—the relationship between wages and labor productivity. From the end of the Second World War through the 1970s, wages and labor productivity grew hand in hand. But by the end of the 1970s something changed and wages grew more slowly compared to increases in labor productivity.

At the same time we have seen a substantial increase in income inequality—so why aren't we still growing together as in the past? Part of the explanation is the marked decline in union coverage over this same period of time. In 1977 more than a quarter of all workers were covered by a collective bargaining agreement—today it is just 12.4 percent.

At the same time, as David Weil (2014) describes in his new book The Fissured Workplace, there has been a large increase in the use of temporary hires, sub-contracting, and third party managers that challenges the regulation of working conditions and setting of wages within a firm. As the distance between employers and their employees has increased, the gains in profitability are shared amongst a smaller number of core workers rather than being more widely distributed. Add to this the threat of international competition, the growing practice of global crowdsourced microtasking such as Amazon's Mechanical Turk, and the deepest recession since the Great Depression, and one can understand why workers are feeling more vulnerable and less likely to successfully negotiate for shared gains.

So what to do? The good news is that for all the political rancor in Washington right now there does seem to be wide agreement that we need more jobs! The disagreement is how to do this. Concerns with the fiscal deficit have meant many legislators are reluctant to pursue additional fiscal stimulus such as was done in the Great Depression. I believe this approach ignores other deficits in this country that undermine our capacity to successfully grow in a way that more widely distributes the benefits of that growth. These other critical deficits include physical and human capital. Failure to address these deficits undermines our ability to address our fiscal challenges associated with an aging population. At a time when the returns to education are so high it makes no sense to be laying off school teachers. And yet this is what we have been doing the last four years. We cannot expect our future workforce to be best in class when the financial support for public higher education is decreased making the cost of attending college prohibitively expensive. But this is what we have done.

Fixing access to schooling only solves part of our human capital deficit. We also need to find a way to raise the skills of those already in the labor market. But who provides this training? My prior work on employer provided training (see Lynch 1991, 1992, 1993, 1994 and Lynch and Black (1998)) showed that workers with less education or who were working in small firms were much less likely to receive employer provided training. With the rise of subcontracting, workers that may have received company provided training in the past in a large vertically integrated firm are less likely to receive this in a firm that specializes in specific occupations—say custodial or maintenance service. This gives rise to a market failure in the provision of worker training and a role for government to fill this gap. For all the cynicism about government training programs we know that training can work. But it must be connected with employer needs and bring workers' skills in reading, writing, math and problem solving up to a standard where they can read basic manuals, write a short note, and add, subtract, multiply and divide. Recent work on the skills gaps in the manufacturing sector by Andrew Weaver and Paul Osterman (2013) argues that these skills are something that someone with a good high school education or some college should have but unfortunately too many workers are missing these skills. The good news is that this is not an insurmountable skills hurtle but resources need to spent to address this gap.

Not only are we under-investing in human capital we are also under-investing in physical capital–in particular the infrastructure that supports our economy. Our roads, rail, airports, ports, telecommunications, water, sewer, and energy efficiency and sustainability are not up to the standards of many other parts of the world[3]. Investing in this infrastructure would bring us into the 21st century and support the private sector's growth and capacity to compete. It would also provide much needed jobs.

Innovation has been a signature feature of our economy but this requires ongoing investment in R&D—in both the private sector and the public—most notably our national labs, research universities, and public support for research. Again, we are cutting while others invest.

Other advanced economies have faced the same technological and trade shocks that we have, but they have not experienced the same increases in inequality or even unemployment. Part of this is due to due to more protection for workers through union coverage and/or employment regulations. As David Weil argued in his keynote address to this conference on Thursday, we need to enforce our current laws but bring a new approach to enforcement that will have a more effective impact on protecting workers.

But doing all this will only go part of the way to renewing shared prosperity in this country. More broadly we need to find a way to pull together rather than pulling apart. As Angus Deaton (2013) and Thomas Piketty (2014) have argued in their recent books, rising inequality and increased concentration of wealth runs the risk of corrupting our democracy. Those with financial means are able to capture politics and their interests drive policy—in directions such as blocking increases in the earned income tax credit or minimum wage, or by reducing investments in public education, government provided training, unemployment insurance, health, and social security. All of these actions would have a disproportionally negative impact the poor and middle class alike.

This is where an organization like LERA can and must play an important role. As the only organization in the country where labor, management, government, and academics, advocates and neutrals come together to discuss and learn about all aspects of labor and employment relations network, we are an important venue for all the stakeholders to come together and work to advance a more inclusive economy. We need to expand the convening role that LERA can play at the local chapter level and nationally with the social partners. As our colleagues in the Tennessee LERA chapter say, "Let's be the place where we shed light rather than heat on the pressing matters facing us today."

Let me finish on an optimistic note—when people talk about returning to the good old days of the 1960s and 1970s I do not join in that nostalgia. While there is still much more that needs to be done to provide equal access and opportunities there have been many changes in the labor market over my working life that have resulted in many more opportunities for women and minorities. As we celebrate the 50th anniversary of the passage of the Civil rights Act, we have seen a significant expansion of protection of the rights of the LGBT community, older workers, and those with disabilities. This has transformed economic opportunities and advanced social inclusion for many.

As I look at our daughter, now in college, I wonder what changes she will see in the labor market over the next 35 years. I hope that the gains in access and opportunity that I experienced thus far over my working life will continue to improve over hers. I hope that she and all other young people in this room do not need to hide behind the potted plants because they do not have the financial means to sit at the table. Most importantly I hope that she sees a renewed commitment in this country to ensuring that the gains in growth and productivity are more broadly shared.

Endnotes

[1] Ray Marshall (1978). "Selective Employment Policies to Achieve Full Employment," Presidential address, Industrial and Labor Relations Research 30th Annual Proceedings, pp 1-9.

[2] See the Bureau of Labor Statistics (2014) American Time Use Survey Table A-7 "Time spent in primary activities by married mothers and fathers in households with children under 18 average for the combined years 2007-11."

[3] See for example the American Society of Civil Engineers *2013 Report Card for America's Infrastructure.*

References

Albanesi, Stefania and Claudia Olivetti. (2014). "Maternal Health and the Baby Boom," *Quantitative Economics*, July, Vol. 5 (2).

American Society of Civil Engineers 2013 Report Card for America's Infrastructure retrieved from http://www.infrastructurereportcard.org/a/documents/2013-Report-Card.pdf

Averett, Susan, Elizabeth H. Peters, and Donald Waldman. (1997). "Tax Credits, Labor Supply and Child Care," *Review of Economics and Statistics*, vol. 79, issue 1, pp. 125-135.

Blau, Francine and Lawrence Kahn. (2013). "Female Labor Supply: Why is the US Falling Behind?" National Bureau of Economic Research working paper 18702, January.

Bureau of Labor Statistics. (2014) American Time Use Survey, Table A-7 retrieved from http://www.bls.gov/tus/tables.htm on July 11, 2014.

Deaton, Angus. (2013). *The Great Escape: Health, Wealth and the Origins of Inequality*. Princeton, N.J.: Princeton University Press.

Fernández, Raquel, Alessandra Fogli, and Claudia Olivetti. (2004). "Mothers and Sons: Preference Formation and Female Labor Force Dynamics," *The Quarterly Journal of Economics*, vol. 119(4), pp. 1249-1299.

Greenwood, Jeremy, Ananth Seshadri and Mehmet Yorukoglu, (2005). "Engines of Liberation," *Review of Economic Studies*, vol. 72(1), pages 109-133.

Goldin, Claudia. (2006). "The Quiet Revolution That Transformed Women's Employment, Education, and Family," *American Economic Review*, vol. 96 (May), 1-21.

Goldin, Claudia and Lawrence F. Katz. (2002). "The Power of the Pill: Contraceptives and Women's Career and Marriage Decisions," *Journal of Political Economy*, vol. 110(4,Aug), 730-770.

Goldin, Claudia, Lawrence F. Katz, and Ilyana Kuziemko, (2006). "The Homecoming of American College Women: The Reversal of the College Gender Gap," *Journal of Economic Perspectives*, vol. 20(4), pages 133-156,

Imbens, Guido and Lynch, Lisa M. (2006). "Re-employment Probabilities Over the Business Cycle," *Portuguese Economic Journal*, vol. 5(2), pages 111-134.

Lynch, Lisa M., (1994). Editor. *Training and the Private Sector: International Comparisons*. Chicago: University of Chicago Press.

Lynch, Lisa M. (1993) "The Economics of Youth Training in the United States," The Economic Journal, November, pp. F97-F116.

Lynch, Lisa M. (1992). "Private Sector Training and its Impact on the Earnings of Young Workers," *American Economic Review*, March, pp. 299-312.

Lynch, Lisa M. (1991). "The Role of Off-the-Job versus On-the-Job Training for the Job Mobility of Women Workers," *American Economic Review*, May pp. 151-156.

Lynch, Lisa M. (1989). "The Youth Labor Market in the 80's: Determinants of Re-employment Probabilities for Young Men and Women," *Review of Economics and Statistics*, Feb. pp. 37-45.

Lynch, Lisa M. (1985) "State Dependency in Youth Unemployment: A Lost Generation?" *Journal of Econometrics*, April, pp. 71-84.

Lynch, Lisa M. (1983). "Job Search and Youth Unemployment," *Oxford Economic Papers*, October, pp. 271-282.

Lynch, Lisa M. and Sandra E. Black. (1998). "Beyond the Incidence of Training: Evidence from a National Employers' Survey," *Industrial and Labor Relations Review*, Oct. pp. 64- 81.

Malveaux, Julianne. (1994). "Tilting Against the Wind: Reflections on the Life and Work of Phyllis Ann Wallace," *The American Economic Review Papers and Proceedings*, Vol. 84, No. 2, May, pp. 93-97.

Marshall, Ray. (1978). "Selective Employment Policies to Achieve Full Employment," Presidential address, Industrial and Labor Relations Research 30th Annual Proceedings, pp. 1-9.

Peters, Elizabeth H. (1986). "Marriage and Divorce: Informational Constraints and Private Contracting," *American Economic Review*, vol. 76(3), pp. 437-54.

Piketty, Thomas. (2014). *Capital in the Twenty-First Century*. Cambridge, MA: Belknap Press.

Sum, Andrew and Ishwar Khatiwada. (2004). "Still Young, Restless, and Jobless: The Growing Employment Malaise Among U.S. Teens and Young Adults," mimeo prepared for American Youth Policy Forum, retrieved from https://repository.library.northeastern.edu/downloads/neu:m0408q35v?datastream_id=content

Weaver, Andrew and Paul Osterman. (2013). "Skill Demands and Mismatch in U.S. Manufacturing: Evidence and Implications," mimeo, Sloan School of Management, MIT and retrieved from https://www.aeaweb.org/aea/2014conference/program/retrieve.php?pdfid=316

Weil, David. (2014). *The Fissured Economy: Why Work Became So Bad for So Many and What Can Be Done to Improve It*. Cambridge, MA: Harvard University Press.

II. 2015 Presidential Address

American Manufacturing, Jobs, and Dignity

MARTIN J. MULLOY
Ford Motor Co.

(Editor's note: Each year the outgoing LERA president gives a speech at the annual meeting on what he or she deems the most important labor and employment issues and what the response to its challenges — from labor, management, academics, practitioners and government — should be from both policy and practice points of view. The following is an edited version of the speech past LERA president Marty Mulloy, Ford Motor Co. vice president for labor affairs, delivered at the Presidential Luncheon on May 30, 2015. The graphics are those that he used in the speech.)

When I first prepared this speech about four months ago. I liked it, thought it was pretty good. But there's so much happening in the world. So, I ripped it up and started again … From the beginning: I'm from a steel-worker family in Gary, Indiana. Growing up, we talked about standing on the shoulders of our fathers. My dad's a mill worker — 45 years — a factory worker. I've got eight brothers. We all worked in the mill.

What I believe passionately, what I care about, is manufacturing in America. It was the road to take myself from a laborer to semi-skilled to a skilled tradesman to a manager to become vice president of a global company. Now, who would dream that? But it wouldn't have happened unless my dad worked at U.S. Steel and unless he was represented by a union. So, this is not Ford Motor Co. talking today. These are my comments.

What I want to talk about today is the Trans-Pacific Partnership or TPP. *(Editor's note: The TPP is a proposed trade agreement among 12 Pacific countries, Australia, Mexico, Canada, Japan, Peru, et al. including the United States, but, significantly, not China).* TPP's stated goal is to lower barriers, such as tariffs, to trade. The legislation currently under consideration in congress is supported by Republicans, led by Senate majority leader Mitch McConnell, and by President Barack Obama. U.S. organized labor is fiercely opposed to the pact, characterizing it as the bigger, meaner brother of NAFTA, which in labor's view has decimated America's manufacturing and the good jobs that go with it.)

How do we know what the value in the auto industry and what it means to our nation and its people? More than 250,000 people work in the U.S. automotive industry. Nearly 2.5 million — one in seventeen working Americans — depend upon the auto industry. This is massive.

I have been just about every country in the world that has an automotive plant. And that's why every country in the world defends its automotive industry. The multiplier effect of an auto job assembly plant is about one to ten. So, for every auto assembler, you have ten other people whose work depends on the automobile industry.

Automakers are the largest exporter in America. Can you believe that? Do you know the company that's the number-one exporter? It's Ford Motor Co. I'm very proud of that.

We had United Steelworkers president Leo Gerard here today. He said 350,000 of his members are dependent upon the automotive industry. And it's not just steel that goes into cars. There's rubber, glass, computer chips.

Sometimes, in my neighborhood, outside Detroit, my neighbors and I talk about factories closing and the human cost. And they're sympathetic, but a most of them are white-collar workers. So they don't think the loss of an auto plant or the decline of American manufacturing affects them.

But losing our industrial base is not just a blue-collar problem. One out of ten corporate scientists and engineers in the United States is in the automotive industry. This loss of our industrial base is an America problem. This is a problem for people matriculating in a variety of occupations in America.

It's also a problem in terms of research and development. For instance, take a look at automotive R&D: Toyota, General Motors, Ford Motor Co. — each spends nine times more in R&D than Exxon. This is how important this industry is. We-I compete against General Motors; I complete against Chrysler, but I consider them my partners because they're providing jobs here in the United States.

The economic impact of the Big Three auto makers is $17.5 billion dollars a year globally with approximately $14 billion spent in the United States. This isn't just slapping steel together and assembling a car. The auto industry is critical to our success as a nation. We're talking about our nation's intellectual and human capital. If we lose the automotive industry, we lose our scientists and engineers. And these are the people that come up with new innovative ideas, technology, and products.

In Michigan, about eighty cents of each dollar spent by Ford, General Motors, and Chrysler is invested in the United States. Thank God President Obama stood up and saved the automotive industry. We didn't go bankrupt at Ford. We're very, very proud of it. *(Editor's note: GM and Chrysler did go bankrupt and were restructured with the federal government's help.)* If Obama had let things go and not intervened, all of the Big Three American auto makers would have liquidated. And the supply base would have liquidated. We would have all gone under. So, we had the right president at the right time step in and save the automotive industry.

In 1985, there were two non-domestic auto plants in the U.S. — in Smyra, Ohio, and Marysville, Tennessee. Today there are 17 assembly plants. Not power train, not stamping, just assembly plants. Only one plant is unionized. We're talking about 100,000 non-unionized workers. If you think it's important to have organized labor in America, think about the good jobs these plants are costing our nation and our citizens who need good jobs.

The next item I want to talk to is about how do we know when the industry's at risk. First of all, let's take a look at NAFTA *(Editor's note: The North American Free Trade Act, passed during the Clinton administration).*

Let me step back and say something. I've been to Mexico many times, and I love the Mexican people. There are no better workers I've met in my life than the Mexicans. If you ask who wants to volunteer to do a job — the Mexican hands are up immediately. So, this is nothing against the Mexicans; they're tremendous people.

But what started off as maquiladoras *(Editor's note: maquiladoras are manufacturing operations in free-trade zone that import material and equipment duty- and tariff-free, then perform assembly and manufacturing operations and ship processed/manufactured products to other countries, sometimes the countries of origin of the original imported material.)* on the U.S.-Mexican border. The maquiladoras have turned into major assembly plants. There are about one hundred thousand jobs there and another one million jobs associated with these jobs.

There are many more plants planned. Where do you think the engineers, the scientists, the lawyers, the engineers, the workers are going to be located? God bless Mexico. But the AFL-CIO has estimated we've lost 700,000 American auto-worker jobs to Mexico since the passage of NAFTA.

So, you look at what's going on in our cities. The police violence against blacks and the subsequent riots in Baltimore just ripped your heart out. But before any of this happened, Baltimore had lost its manufacturing base. I'm from Gary, Indiana. Go to Gary, Indiana. Go to the west side of Chicago. Go to the south side of Chicago — that used to be the manufacturing hub of the United States. No more. We've lost those jobs.

Here's what happened.

NAFTA was signed in '94. Take a look at today's U.S. trade deficit — 175 billion dollars. This represents an exponential job loss. The way it happens is once you have auto-assembly plants up and running, you get what's called "bulls-eye sourcing". Parts and equipment suppliers — and the jobs they provide — follow. It isn't instant. It takes a while. There has to critical mass of assembly plants for the suppliers to follow.

Then there was the Korea Free-Trade Act passed 2012. Again, God bless the Koreans, hard-working people. But after the free-trade act, the U.S. trade deficit with Korea increased fifty percent. We lost more than 50,000 jobs. Imported vehicles and parts increased by 31 percent. U.S. automotive imports from Korea went up two-hundred and ninety-four percent, and our U.S. auto deficit increased 32 percent. This is very predictable. Our trade imbalance is increasing every year, and it's going continue to increase.

The average annual market for cars in North America's is seventeen million units. Korea's is less than one million. The free-trade act opens up our market with no tariffs. So, the ratio of market opportunity is seventeen to one. At the end of the day, they got a way better deal. For all those reasons, it's tough to take.

And, last, the Trans-Pacific Partnership — the TPP. And God bless the Japanese, great people. But they are going to act in their own interest for their population, for their people. They're acting no different than I would if I were in charge of Japan.

The total value of the economies involved in TPP is 28 trillion dollars. This is 39 percent of the global gross domestic product. The U.S. GDP is 17.5 trillion dollars. Japan's GDP is 4.6 trillion dollars. So this adds up to 21.6 trillion dollars of the 28 trillion dollar global GDP total. So this pact is really mainly about trade between the U.S. and Japan.

Ford's biggest problem with the pact is currency manipulation. The short explanation of currency manipulation is that a country sells its own currency and buys foreign currency. This reduces the demand for its currency and lowers the value to gain an unfair competitor advantage. The way it works out is that the Japanese get an advantage of $6,000 per vehicle over our comparative U.S.-made vehicles. That's $6,000 multiplied by every vehicle sold. It's billions and billions of dollars. And at the end of the day, we are going to have a big loss of American jobs.

One American car is sold in Japan for each one hundred and thirty Japanese cars sold here.

In the U.S., 59 percent of cars are made here; 41 percent are foreign made. In Japan, 94 percent of the cars are made there. We just can't get into that market. This is why the U.S. has a 52 billion dollar trade deficit with Japan. It's an unfair fight. In these trade deals, we're like the kid who gets his lunch money stolen every morning.

Japan has manufacturing capacity for 11 million units; they produce 8.9 million units annual. And the sales within Japan are 4.8 million units. So, if you just roughly say they produce about 400,000 to 450,000 vehicles per plant, Japanese domestic consumption requires about 10 plants. Exports require about nine plants. So the Japanese now have five to seven plants sitting completely or mostly idle. They're not being used, or if they're being used, their output is so low that they ought to be closed.

Our argument is that we're trading our auto-manufacturing capacity here for loading up the unused Japanese capacity. It's pretty simple. This means there is going to be less production of U.S.-made cars here and more production of Japan's vehicles in Japan. So, they're going to hire more people. The question is are we going to make that bad tradeoff? Remember, the auto industry is one of the key segments of the entire U.S. economy.

And that's how you get the growing trade deficit. More than 70 percent of the trade deficit with Japan is automotive. And it adds up to more than 50 billion dollars a year. And you say politics makes strange bedfellows? I just don't understand. McConnell and Obama can't get along on anything, and yet they're on the same side on TPP.

The big question in the TPP is how open the countries' markets are. The United Kingdom is more than 80 percent open, Spain about 70 percent open, the U.S. and Germany about 40 percent open. We can't get into the Japanese and Korean markets. Their markets are effectively closed to us, so it's very difficult.

As the Ford Motor Co. vice president of labor affairs, I went through some horrific times. I was involved in either closing or selling 34 plants — 11 assembly plants. Assembly plants are aircraft carriers. Each one is a whole city. When you shut down an assembly plant, you wipe out an entire community.

And restructuring is even worse — the most difficult thing I've ever done in my life. I'd sit across from Ron Gettelfinger and Bob King of the UAW and explain to them we're going have to close another plant because we're going through a recession, a depression. If we don't close those plants and get demand equal to capacity, we're going to go out of business.

And lastly, what are the implications for labor and management going forward? We lost 17 billion dollars in 2006 before Lehman Brothers collapsed in 2008. We used to quote James Stockdale: "You must never confuse faith that you will prevail in the end with the discipline to confront the most brutal facts of your current reality, whatever they may be." This is what we said during those toughest, most gut-wrenching times of Ford Motor Co.

I don't know how many of you remember Stockdale. He was the commanding officer in the Hanoi Hilton in the waning days of the Vietnam War. He was the guy who kept those soldiers alive.

You heard Ron Bloom talk about the importance of organizations getting out of denial, in our case that we were going to go bankrupt. *(Editor's note: Bloom, who spoke at the meeting, was assistant for manufacturing policy from 2009-11 for President Barack Obama. He was also instrumental in restructuring of the American automobile industry following the bankruptcies of GM and Chrysler.)* Those were tough meetings, sitting down with the UAW and taking a look at the facts. At the end of the day, we went from making about 4.2 million units to 1.8 million units. That's why we closed so many plants.

And that's why we had 50,000 buyouts of our workers. Of those 50,000, there was not one involuntary buyout. Not one. And we made sure our people got a good landing. We had something called an educational-opportunity program — 15,000 dollars a year for college tuition for four years. Our former workers got half-pay, full medical benefits, and a 50,000 dollar life insurance policy for their families in case they died. We offered that deal to every one of those 50,000 employees.

At Ford, it's management and labor together. You can't tell the hourly from the salaried people. People think negotiations are between me with my CEO and CFO sitting across the table from the UAW's Bob King and his negotiating team. I would always say, "Guys, we're all in this boat together. We're going to sink or swim together."

People think there are two sides to a negotiation. There are three. You've got to bring them — you, me and us — all together. I learned that from Joel Cutcher-Gershenfeld (longtime Ford consultant, past LERA president and Brandeis professor).

There are two of our hourly workers in the audience today—Dan Brooks. Dan was the UAW guy who handled safety and quality. Honestly, sometimes he drove me nuts. But he had the intellect, drive and ambition to hold management accountable to make sure our quality and safety practice were the best. We worked together. Dan and Joel and I wrote a book (Inside the Ford-UAW Transformation, MIT Press) together.

Also in the audience is Armentha Young, a Six Sigma black belt from the Ford Dearborn truck plant. *(Editor's note: The Six Sigma black belt certifies that that a worker has mastered a disciplined, data-driven approach and methodology for eliminating defects in manufacturing, product and service.)* It takes most people several tries to pass the test for the black belt. Armetha? She only had to take the test once.

Now, I want to show you a picture. That's an F-150, all-aluminum body truck. (*Editor's note: The F-150 had been the best-selling truck in the world for years when Ford undertook this radical re-engineering and re-design.)* This is the highest-tech vehicle on the road. Everyone's going copy us. It's got a turbo-boost engine. By the way, wherever I speak, I want to sell one Ford product.

And by the way, if you believe in unions, and you believe in union and company working together, and you're not driving a union vehicle, shame on you.

I want to tell you a story. A gentleman from the NAACP came to see me. He came into my office, and he said "Thank you."

"What did I do for you?" I asked.

He said: "Do you know where the black middle class in America comes from? It comes from your factories. Where do the manufacturing people in the black community work? They work at Ford, General Motors, and Chrysler. You weren't afraid to invest in the urban areas. You're not afraid to promote the people. You're not afraid to provide educational opportunities."

When I was watching the problems in Baltimore on TV and another NAACP fellow talking about how the fathers mostly weren't around and if they were around, they weren't working. There are no jobs.

Our country has got to come together and provide opportunity for people. With work comes dignity. This is why I find TPP so painful. The train is coming. And it's going to wreck. We're going to lose key manufacturing jobs.

That's why I picked this topic — because I had one shot to talk to influential people before the decision making. Go home, talk to your congressman, talk to your senator. Get your voice out there. And let's work together. Because I grew up in a union household. My uncles, my aunts, my dads, my brothers, they're all union. I just happened to end up on the other side. I wanted to work for the union, but they wouldn't take me.

Again, thank you very much, and everyone have a safe trip home.

III. The Employer Ownership Approach to Shared Prosperity

Employee Ownership and Firm Survival Through the Great Recession

FIDAN ANA KURTULUS
University of Massachusetts Amherst

DOUGLAS KRUSE
Rutgers University

Abstract

This study empirically examines the relationship between employee ownership and firm survival through the Great Recession. Using a dataset comprised of all publicly traded companies in the United States during 1999 to 2010, the analysis shows that firms with employee ownership programs had a higher likelihood of survival than firms without employee ownership, and this result is robust to controlling for firm size, industry and union status. The analysis considers a range of different variables measuring the incidence of employee ownership. Our findings have important policy implications about the potential of employee ownership for lowering economy-wide job loss and unemployment.

Introduction

This study examines the relationship between employee ownership and firm survival among publicly traded companies in the United States during the period 1999-2010. Employee ownership refers to situations where workers have an ownership stake in the company where they work, for example through ownership of company stock and stock options. There is a fairly large literature showing that employee ownership has positive impacts on numerous firm and worker outcomes. Employee ownership enables workers to partake in the financial results of their companies, and this makes workers more committed to the firm and more closely aligns the interests of the worker with those of the firm, leading to improved worker and firm performance. Giving workers a stake in the success of the firm can motivate higher levels of effort, generate greater job satisfaction, and induce greater cooperation among co-workers and between workers and management.

Perhaps unsurprising given the numerous benefits of employee ownership, the prevalence of broad based employee ownership has been steadily growing in the United States as well as in other developed economies (Blasi, Freeman and Kruse 2013, Hashi and Hashani 2013). According to a 2013 study, nearly half of private-sector full-time wage and salary workers have some form of share in the firm where they work, about 21 percent of workers have employee stock ownership in the company for which they worker, and about 10 percent report having stock options in their employer. (Kruse, Freeman, and Blasi 2010). Many of the most successful companies in the United States offer broad based employee ownership, including Fortune 500 companies and large multinationals like Google, Exxon Mobil, Amazon, Procter and Gamble, and Ford Motor Company (Blasi, Kruse and Freeman 2013).

This paper specifically examines the relationship between employee ownership and firm survival among publicly traded firms in the United States during the period 1999-2010. Firm survival is an important outcome variable to examine as it is generally an indicator of success for a company, increases job security for workers employed, and thereby benefits the economy more broadly by reducing unemployment and economic hardship.

What are the channels through which employee ownership may enhance firm survival? First, past research has shown employee ownership to be linked to increased firm and worker productivity through greater employee cooperation, commitment and information sharing at the workplace. Despite the free-rider problem associated with group-based pay, studies have generally found employee ownership to be linked to higher performance (Kaarsemaker 2006, Doucialiagos 1995, Kruse and Blasi 1997). Second, prior research has shown that employee ownership policies tend to be implemented along with other complementary high-performance workplace practices, such as employee involvement in decision making, team production, and on the job training, to create a more engaged workplace, which can also contribute to improved survival outcomes in the face of financial distress (Ichniowski et. al. 1996, Ichniowski, Shaw and Prennushi 1997, Becker and Huselid 1998, Blasi, Freeman, Mackin and Kruse 2010).[1] Third, employee ownership can reduce workplace conflict that can contribute to production frictions and firm failure (Cramton et. al. 2008). Fourth, employee ownership tends to foster long-term employment relationships, which in turn encourage both employers and employees to make higher investments in firm-specific skills and facilitate increased productivity and survival prospects (Levine and Parkin 1994). Fifth, employee ownership may increase employee willingness to suggest and participate in innovations that enhance long-term firm performance and prospects (Kruse, Freeman, and Blasi 2010). On the other hand, employee ownership may negatively impact firm survival if it contributes to increased communication frictions among employees, or otherwise creates conflicts or inefficiencies within firms that can lead to a higher rate of failure (Hansmann 1996).

In this study, we use data on the entire universe of publicly traded U.S. companies as of 1999, following them through 2010. We use both federal Form 5500 data, which is collected by the Department of Labor and provides information on employee participation in employee ownership through pension plans, and Standard and Poor's CompuStat data, which provides information on firm closures, mergers, acquisitions, liquidations, as well as firm financial data. We estimate Cox hazard regressions predicting the probability of firm dissolution as a function of a wide array of measures of employee ownership and firm controls. Our results generally cast doubt on theoretical concerns about possible inefficiencies of employee ownership, and instead support the idea that employee ownership improves the likelihood of firm survival. If employee ownership does improve survival this can have favorable effects on the economy as a whole, which can justify supportive public policy as discussed in the conclusion.

Empirical Methodology

To conduct our empirical analysis of the link between employee ownership and firm survival, we compiled a dataset by merging Standard and Poor's Industrial CompuStat database on publicly-traded firms and the Form 5500 pension plan data collected by the U.S. Department of Labor for the years 1999 to 2010. The CompuStat database provides information on firm characteristics like total employment, industry, financial information, and reason for firm failure, while the Form 5500 pension plan database contains detailed information on employee ownership in defined contribution plans and Employee Stock Option Plans (ESOPs). We matched firm records from the CompuStat data and Form 5500 data using each firm's unique IRS Employer Identification Number.

Our dataset is comprised of the full sample of publicly-traded companies in the U.S., which is an important improvement over datasets drawn from special surveys suffering from small sample sizes and bias from self-selection of respondents. A further advantage is the twelve year span of our dataset, covering a decade when the U.S. experienced two recessions, in 2001 and 2008, allowing us to examine how employee ownership firms weathered these economic downturns relative to non-employee ownership companies. We also have a rich array of measures of employee ownership at companies, including the presence of employee ownership through pension programs and ESOPs, and the extent of such employee ownership in terms of total participation and the share of the firm owned by workers.

We estimate Cox proportional hazards regressions to predict the likelihood of firm failure. The Cox proportional hazards model specifies the hazard rate (i.e., the probability of failure) for firm j as a function of firm characteristics. The main firm characteristic of interest in our hazards models is employee ownership. Our hypothesis is that in our Cox proportional hazards regressions the relative hazard ratio for this variable should be between zero and one, indicating a lower likelihood of failure for EO firms than non-EO firms on average. Our regressions also include firm controls, including firm size, union status, and industry.

We first estimate regressions where we treat any disappearance of a firm from the CompuStat database as a firm failure. However, companies may disappear as independent entities when they merge or are acquired by another company, and this can actually signal success in some cases, as other firms want to acquire or merge with successful companies. CompuStat provides reason for deletion of firms which no longer appear in that database, including acquisition, merger, bankruptcy and liquidation. We therefore also estimate models in which firm failure is defined strictly as bankruptcy or liquidation.

We consider a range of different measures of employee ownership in our empirical analyses. To start, we consider whether the firm reported any employee ownership stock in any of its defined contribution pension plans, including employee ownership in 401(k) Plans, Employee Stock Ownership Plans (ESOPs), and deferred profit sharing plans. We then go on to examine wither the firm reported any Employee Stock Ownership Plans (ESOPs). Third, we consider the dollar value of employee ownership stock per employee at the firm (in $1000s). Finally, we consider whether the share of the firm owned by employees is greater than or equal to 5% (meeting the Security and Exchange Commission's definition of a major stockholder).

Descriptive Statistics

We now turn to an analysis of the general attributes of the firms in our database. Table 1 (all tables in this paper are at the end) shows summary statistics for all variables in our pooled sample. On average, firms had 6,068 employees. The probability that a firm disappeared from the sample was 7.2 percent, and the probability of bankruptcy or liquidation was 0.4 percent, on average. The majority of firms in the sample were in the manufacturing industry, comprising 36 percent of firms, followed by finance and services, comprising 22 and 18 percent, respectively. 19 percent of firms reported employee ownership stock in their defined contribution plans, and 8 percent had ESOPs in particular.

Table 2 depicts descriptive statistics by presence of employee ownership programs in the pooled sample, and illustrates, among other results, that EO firms were on average larger, were less likely to face bankruptcy/liquidation and generally less likely to disappear than non-EO firms. Employee ownership firms were also more likely to be in the transportation and finance industries while less likely to be in the service industry, than non-EO firms. On average the dollar value of employee ownership stock per employee was $12,194, and the share of a firm owned by employees was 3 percent.

Figure 1 reflects the main story of this study. It illustrates failure rates through 2010 among firms that were observed in 1999 and their employee ownership status in 1999. The two lines in the figure represent the share of 1999 firms that disappeared in each ensuing year, by whether the share of the firm owned by employees is above or below five percent in 1999.[2,3] As we can see, the share of firms that disappeared was lower among firms where the share of the firm owned by employees is above five percent. This figure depicts the basic story that employee ownership and firm survival are positively related, however, the numbers reflect uncontrolled mean survival rates. We next turn to investigating whether this story holds up when we control for firm characteristics such as size and industry.

Figure 1
Failure Rates by Share of Firm Owned by Employees >=5%
vs. Share of Firm Owned by Employees <5%

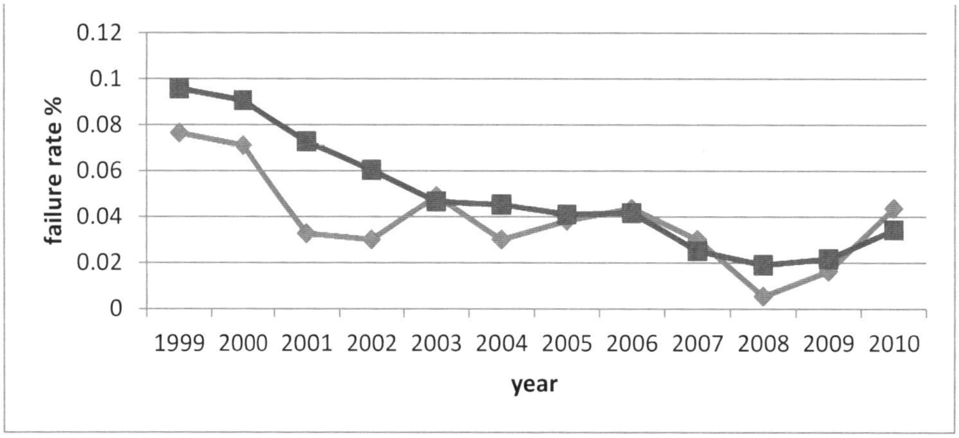

Note: Tracks the share of 366 firms with Share of Firm Owned by Employees>=5% (diamonds) and 9,542 firms with Share of Firm Owned by Employees<=5% (squares) observed in 1999 that were no longer observed in ensuing years.

Does Employee Ownership Enhance Firm Survival?

We now turn our attention to regression analysis of the relationship between employee ownership and firm survival. Table 3 presents hazard ratios from Cox proportional hazard regressions predicting the likelihood of firm death. The specification in the first column considers firm disappearance for any reason, including mergers and acquisitions, while the specification in the second column defines firm failure strictly as bankruptcy or liquation.

Panel A illustrates the results for ANYEO indicating whether a firm reported any employee ownership stock in any of its defined contribution pension plans, and indicates that EO firms were less likely to disappear than non-EO firms. As seen in the first column, the relative hazard ratio for ANYEO is 0.786 and statistically significant at the one percent level, meaning that EO firms were only 78.6 percent as likely as non-EO companies to disappear in any year over the 1999-2010 period, controlling for firm size, union status and industry. Turning to the second column using the more restricted measure of failure, we see that EO firms were only 92.8 percent as likely as non-EO companies to experience bankruptcy or liquation in any year over this period. The hazard ratio in the second column is not statistically significant at conventional levels, however this is due at least in part to the much smaller sample size in this restricted model--there are only 303 failures as opposed to over 6000 in the specification in the first column. The results continue to indicate an economically significant relationship to firm failure.

Panel B shows the hazard ratios are similar using ESOPI indicating whether a firm reported any Employee Stock Ownership Plans (ESOPs). Firms with ESOPS are 82.1 percent as likely as non-ESOP firms to disappear in any year. The estimates for bankruptcy or liquidation indicate ESOP firms are 90 percent as likely to become bankrupt or liquidated, however, as before, although this result is economically significant it is not statistically significant at conventional levels.

As seen in Panel C, the value of EO stock per worker (EOEEI) is associated with a higher survival probability: As seen in the first column, an extra $1000 of employee ownership stock is linked to a 0.5% lower risk of disappearing and this result is statistically significant at the five percent level. This means that, since the mean value of employee ownership stock among employee owners is $12,194, average employee ownership is linked to a 6.097 percent lower risk of disappearing in any given year. The relative hazard ratio for the more stringent firm death measure (bankruptcy and liquidation) is very similar in magnitude however does not achieve statistical significance at conventional levels due to the smaller sample size.

The share of the firm owned by employees had a big impact on firm survival: as seen in Panel D, firms where the share of the firm owned by employees was five percent or more were only 77.2 % as likely to disappear than firms with less than a five percent share of employee ownership. The estimate for the more stringent firm death measure (bankruptcy and liquation) indicates that firms where the share of the firm owned by employees was five percent or more were only 81.3 percent as likely to face bankruptcy or liquidation, but this results is not statistically significant, as before due to the small sample size.

Causation or Correlation?

A common question in most data-driven studies is whether the relationship unearthed is a causal association or simply a correlation. It is important to note that the empirical analysis above does not identify a causal relationship, but simply an observed correlation between employee ownership and firm survival. The fact that employee ownership firms persist for longer periods of time may partially reflect self-selection of both the firms and the workers employed at the EO firms. For example, it may be that employee ownership firms are more likely to self-select into markets or sectors where firm dissolution is more rare; we have tried to address this concern by including industry controls in our hazard regressions. Additionally, there may be self-selection of workers into EO firms in a way that makes these firms more likely to persist. Workers may self-select by unobservable characteristics that can affect firm survival. As Chiaporri and Salanie (2003) point out, a combination of unobserved heterogeneity and endogenous matching of agents to employment contracts can create selection based on the parameters of interest. For example, if more productive or motivated workers are drawn to work at firms that share wealth their employees through their EO schemes, these firms may be more likely to succeed through turbulent times.

Unfortunately selection is a potential threat to identifying a causal relationship in all studies using non-experimental data (Kremer, 1997). Selection bias is most effectively controlled through random assignment in experiments. While there have not been any experiments in the survival of employee ownership firms, it is worth noting that experiments have been conducted regarding the economic performance effects of group-based incentives, with results indicating causal effects on performance. A randomized field experiment showed favorable effects of firm-based financial incentives on turnover and productivity (Peterson and Luthans, 2006). This is complemented by laboratory experiments conducted by Frohlich et al. (1998) finding higher productivity in researcher-designed "employee-owned firms," and by Mellizo (2013) finding that randomly assigned subjects in shared compensation contracts had higher individual performance than those in individual compensation contracts even in the absence of other high-performance workplace practices like decision-making autonomy. While randomized experiments are hard to imagine on the topic of firm survival, these results call out for innovative research strategies to determine the causal role of employee ownership.

Conclusion

This study finds that publicly traded companies in the U.S. with employee ownership programs were more likely to survive the last two recessions. From a research perspective, this is important because it raises questions about the mechanisms and workplace dynamics underlying employee ownership and firm survival, suggesting that employee ownership may affect firm incentives and policies in a way that enhances firm sustainability. It is possible, for example, that employee ownership combines with employee involvement, job training, and job security to create an "ownership culture" that may not only improve short-term performance, but also contributes to employee commitment and innovative ideas that enhance long-term survival.

From a policy perspective, our finding that firms with employee ownership were more likely to survive these recessions is an important result because higher survival rates among employee ownership companies could lead to lower job loss rates and unemployment. Employee ownership may therefore serve as an economy-wide policy instrument to lower job loss through increasing the likelihood of firm survival. This indicates positive externalities on the overall economy and government expenditures (e.g., for unemployment compensation and social programs) that would justify supportive public policy, given the large economic and social costs of unemployment resulting from firm closings.

Acknowledgments

The authors thank Joseph Blasi, Richard Freeman, Takao Kato, Paula Voos, and seminar participants at the 2015 LERA Meetings and 2015 Kelso Research Conference on Shared Capitalism at Rutgers University for valuable comments and suggestions, and the Upjohn Institute for Employment Research for research funding.

Endnotes

[1] Although it is important to note that recent laboratory experiments conducted by Mellizo (2013) in which subjects are randomly assigned to shared versus individual flat-wage compensation contracts show that shared pay motivates higher individual performance even in the absence of other high-performance workplace practices like decision-making autonomy.

[2] A similar pattern is observed when we compare trajectories by presence of employee ownership in all years during 1999-2010.

[3] These figures pertain to firm failure for any reason; figures for failure strictly due to bankruptcy or liquidation appear similar despite the considerably smaller sample sizes.

References

Becker, Brian E. and Mark A. Huselid. 1998. "High-Performance Work Systems and Firm Performance: A Synthesis of Research and Managerial Implications," in *Research in Personnel and Human Resources*, G. Ferris, Ed., Vol. 16, Greenwich, Connecticut: JAI Press.

Blair, M. M., Kruse, D. L., and Blasi, J.R. (2000). Employee Ownership: An Unstable Form or Stabilizing Force? In: M. Blair and Kochan (Eds.), *The New Relationship: Human Capital in the American Corporation.* Washington, DC: Brookings Institution.

Blasi, Joseph, Richard Freeman and Douglas Kruse. 2013. *The Citizen's Share: Putting Ownership Back Into Democracy.* New Haven: Yale University Press.

Blasi, Joseph, Douglas Kruse and Dan Weltman. 2013. "Firm Survival and Performance in Privately-Held ESOP Companies," *Advances in the Economic Analysis of Participatory and Labor-Managed Firms*, Volume 14 pp.109-124.

Blasi, Joseph R., Richard B. Freeman, Christopher Mackin, and Douglas L. Kruse. 2010. "Creating a Bigger Pie? The Effects of Employee Ownership, Profit Sharing, and Stock Options on Workplace Performance," in *Shared Capitalism at Work: Employee Ownership, Profit and Gain Sharing, and Broad-Based Stock Options*, Douglas L. Kruse, Richard B. Freeman, andJoseph B. Blasi, Eds. Chicago, IL: University of Chicago Press.

Burdin, Gabriel. 2014. "Are Worker-Managed Firms More Likely to Fail than Conventional Enterprises? Evidence from Uruguay." *Industrial and Labor Relations Review*, Vol. 67(1), pp. 202-238.

Chiaporri, Pierre Andre, and Bernard Salanie. 2003. "Testing Contract Theory: A Survey of Some Recent Work." In *Advances in Economics and Econometrics,* Vol. 1. M. Dewatripont, L. Hansen, and S. Turnovsky (eds.). Cambridge University Press.

Cramton, Peter, Hamid Mehran , Joseph S. Tracy. 2008. "ESOP Fables: The Impact of Employee Stock Ownership Plans on Labor Disputes," FRB of New York Staff Report No. 347 Federal Reserve Board of New York.

Doucialiagos, C. 1995. "Worker Participation and Productivity in Labor-Managed and Participatory Capitalist Firms: A Meta-Analysis." *Industrial and Labor Relations Review*, 49(1), 58-77.

Estrin, Saul and Derek Jones. "The Viability of Employee-Owned Firms: Evidence from France." *Industrial and Labor Relations Review*, 45(2), 323-338.

Frohlich, N., Godard, J., Oppenheimer, J.A. and Starke, F. (1998). Employee Versus Conventionally-Owned and Controlled Firms: An Experimental Analysis. *Managerial and Decision Economics*, 19, 311-326.

Hansmann, H. (1996). *The Ownership of Enterprise*. Cambridge, MA: Harvard University Press

Hashi, Iraj and Alban Hashani. 2013. "Determinants of Financial Participation in the EU: Employers' and Employees' Perspective." In *Advances in the Economic Analysis of Participatory and Labor-Managed Firms*, Vol. 14, pp. 187-215. Emerald Publishing.

Ichniowski, Casey, Kathryn Shaw, and Giovanna Prennushi. 1997. "The Effects of Human Resource Management Practices on Productivity: A Study of Steel Finishing Lines," *American Economic Review*, Vol. 87, No. 3, pp. 291-313.

Ichniowski, Casey, Thomas Kochan, David Levine, Craig Olson, and George Strauss. 1996. "What Works at Work: Overview and Assessment," *Industrial Relations*, Vol. 35, No. 3, pp. 299-333.

Kaarsemaker, E. 2006. "Employee Ownership and Its Consequences: Synthesis-Generated Evidence for the Effects of Employee Ownership and Gaps in the Research Literature." York, UK: University of York.

Kremer, Michael. 1997. "Why Are Worker Cooperatives So Rare?." NBER Working Paper 6118.

Kruse, Douglas, Richard Freeman and Joseph Blasi. 2010. *Shared Capitalism at Work: Employee Ownership, Profit and Gain Sharing, and Broad-Based Stock Options*. Chicago: University of Chicago Press.

Kruse, Doutlas and Joseph Blasi. 1997. "Employee Ownership, Employee Attitudes, and Firm Performance: A Review of the Evidence." In D. J. B. Mitchell, D. Lewin, and M. Zaidi (eds)., *Handbook of Human Resource Management*, pp. 113-151. Greenwich, CT: JAI Press.

Levine, David and Richard Parkin 1994. "Work Organization, Employment Security and Macroeconomic Stability." *Journal of Economic Behavior and Organization*, 24, 251-271.

Logue, J. and Yates, J.S. 1999. "Worker Ownership American Style: Pluralism, Participation and Performance." *Economic and Industrial Democracy*, 20, 225-252.

Mellizo, Phil. 2013. "Can Group-Incentives Without Participation Survive the Free-Rider Problem? A View From the Lab," in Douglas Kruse (ed.) *Sharing Ownership, Profits, and Decision-Making in the 21st Century (Advances in the Economic Analysis of Participatory & Labor-Managed Firms, Volume 14)*. Emerald Group Publishing Limited, pp. 27-59.

Park, Rhokeun, Douglas Kruse and James Sesil. 2004. "Does Employee Ownership Enhance Firm Survival?." In *Advances in the Economic Analysis of Participatory and Labor-Managed Firms*, Vol. 8, pp. 3-33.

Perotin, Virginie. 1987. "Conditions of Survival and Closure of French Worker Cooperatives: Some Preliminary Findings." In D. Jones and J. Svejnar (Eds.), *Advances in the Economic Analysis of Participatory and Labor-Managed Firms* (Vol.2). Greenwich, CT: JAI Press.

Perotin, Virginie. 1997. *What Makes Coops Work? Institutional Viability, Firm Creation, Survial and Closure Among Workers' Cooperatives in France*. Ph.D. Dissertation, Cornell University.

Peterson, Suzanne J., and Fred Luthans. 2006. "The Impact of Financial and Nonfinancial Incentives on Business-Unit Outcomes Over Time," *Journal of Applied Psychology*, Vol. 91, No. 1, 156–165.

Table 1
Descriptive Statistics and Variable Definitions

Variable	Definition	Mean	Std. Dev.	Min	Max	Obs
Firm Death	Dummy variable equaling 1 if firm i in year t dropped out of the dataset for any reason (including bankruptcy, merger/acquisition, liquidation, reverse acquisition, no longer publicly traded, no longer files with SEC);0 otherwise	0.072	0.258	0	1	95,855
Strict Firm Death	Dummy variable equaling 1 if firm i in year t went bankrupt or was liquidated; 0 otherwise	0.004	0.060	0	1	95,855
Employees	Number of employees(in 1000s)	6.068	32.860	0	2,100	82,900
Collective Bargaining	Dummy variable equaling 1 if firm i in year t has a collectively bargained pension plan; 0 otherwise	0.077	0.267	0	1	95,855
Firm reported any EO stock	Dummy variable equaling 1if firm i in year t reported any employee ownership stock in any of its defined contribution plans, including employee ownership in 401(k) Plans, Employee Stock Ownership Plans (ESOPs), and deferred profit sharing plans; 0 otherwise	0.188	0.390	0	1	95,855
Firm reported any ESOPs	Dummy variable equaling 1 if firm i in year t reported any Employee Stock Ownership Plans (ESOPs); 0 otherwise	0.084	0.277	0	1	95,855
Value of EO stock per worker	Value of employee ownership stock per employee (all; including employees who do not participate in an employee ownership plan) at firm i in year t	2.242	11.937	0	937.660	95,413
Share of firm owned by employees	Share of firm i owned by employees in year t	0.008	0.031	0	1.038	71,204
Share of firm owned by employees >= 5%	Dummy variable equaling 1 if share of firm i owned by employees in year t is greater than or equal to 5%; 0 otherwise	0.035	0.183	0	1	95,855
Share of firm owned by employees >= 10%	Dummy variable equaling 1 if share of firm i owned by employees in year t is greater than or equal to 10%; 0 otherwise	0.013	0.112	0	1	95,855
Agriculture	Dummy variable equaling 1 if the industry of firm i in year t is Agriculture, Forestry and Fishing; 0 otherwise	0.003	0.056	0	1	95,855
Mining	Dummy variable equaling 1 if the industry of firm i in year t is Mining; 0 otherwise	0.045	0.207	0	1	95,855
Construction	Dummy variable equaling 1 if the industry of firm i in year t is Construction; 0 otherwise	0.009	0.092	0	1	95,855
Manufacturing	Dummy variable equaling 1 if the industry of firm i in year t is Manufacturing; 0 otherwise	0.357	0.479	0	1	95,855
Transportation	Transportation, Communications, Electric, Gas and Dummy variable equaling 1 if the industry of firm i in year t is Sanitary Services; 0 otherwise	0.094	0.292	0	1	95,855
Wholesale Trade	Dummy variable equaling 1 if the industry of firm i in year t is Wholesale Trade; 0 otherwise	0.030	0.170	0	1	95,855
Retail Trade	Dummy variable equaling 1 if the industry of firm i in year t is Retail Trade; 0 otherwise	0.051	0.220	0	1	95,855
Finance	Dummy variable equaling 1 if the industry of firm i in year t is Finance, Insurance and Real Estate; 0 otherwise	0.218	0.413	0	1	95,855
Services	Dummy variable equaling 1 if the industry of firm i in year t is Services; 0 otherwise	0.177	0.382	0	1	95,855
Public Administration	Dummy variable equaling 1 if the industry of firm i in year t is Public Administration; 0 otherwise	0.016	0.125	0	1	95,855

Table 2
Descriptive Statistics By Presence of Employee Ownership

	Firm reported any EO stock = 1			Firm reported any EO stock = 0		
	Mean	Std. Dev.	Obs	Mean	Std. Dev.	Obs
Firm Death	0.052	0.221	17,981	0.076	0.265	77,874
Strict Firm Death	0.002	0.048	17,981	0.004	0.062	77,874
Employees (in 1000's)	13.857	59.312	17,545	3.977	20.115	65,355
Collective Bargaining	0.208	0.406	17,981	0.047	0.212	77,874
Firm reported any EO stock	1	0	17,981	0.000	0.000	77,874
Firm reported any ESOPs	0.350	0.477	17,981	0.022	0.148	77,874
Value of EO stock per worker	12.194	25.571	17,539	0	0	77,874
Share of firm owned by employees	0.034	0.057	16,658	0	0	54,546
Share of firm owned by employees >= 5%	0.186	0.389	17,981	0	0	77,874
Share of firm owned by employees >= 10%	0.068	0.252	17,981	0	0	77,874
Agriculture	0.004	0.060	17,981	0.003	0.055	77,874
Mining	0.034	0.180	17,981	0.048	0.213	77,874
Construction	0.011	0.103	17,981	0.008	0.089	77,874
Manufacturing	0.368	0.482	17,981	0.355	0.478	77,874
Transportation	0.113	0.316	17,981	0.090	0.287	77,874
Wholesale Trade	0.030	0.170	17,981	0.030	0.170	77,874
Retail Trade	0.069	0.254	17,981	0.047	0.211	77,874
Finance	0.254	0.435	17,981	0.210	0.407	77,874
Services	0.116	0.320	17,981	0.191	0.393	77,874
Public Administration	0.003	0.057	17,981	0.019	0.136	77,874

Table 2 (continued)
Descriptive Statistics By Presence of Employee Ownership

	Firm reported any ESOPs = 1			Firm reported any ESOPs = 0		
	Mean	Std. Dev.	Obs	Mean	Std. Dev.	Obs
Firm Death	0.049	0.217	8,027	0.074	0.261	87,828
Strict Firm Death	0.002	0.043	8,027	0.004	0.061	87,828
Employees (in 1000's)	20.991	80.673	7,749	4.529	22.244	75,151
Collective Bargaining	0.287	0.452	8,027	0.058	0.234	87,828
Firm reported any EO stock	0.783	0.412	8,027	0.133	0.340	87,828
Firm reported any ESOPs	1	0	8,027	0	0	87,828
Value of EO stock per worker	18.564	34.216	7,846	0.779	4.937	87,567
Share of firm owned by employees	0.049	0.074	7,360	0.003	0.016	63,844
Share of firm owned by employees >= 5%	0.307	0.461	8,027	0.010	0.099	87,828
Share of firm owned by employees >= 10%	0.122	0.328	8,027	0.003	0.053	87,828
Agriculture	0.006	0.076	8,027	0.003	0.053	87,828
Mining	0.019	0.135	8,027	0.047	0.212	87,828
Construction	0.008	0.090	8,027	0.009	0.092	87,828
Manufacturing	0.312	0.463	8,027	0.361	0.480	87,828
Transportation	0.138	0.345	8,027	0.090	0.287	87,828
Wholesale Trade	0.028	0.164	8,027	0.030	0.170	87,828
Retail Trade	0.052	0.222	8,027	0.051	0.220	87,828
Finance	0.369	0.483	8,027	0.204	0.403	87,828
Services	0.066	0.247	8,027	0.188	0.390	87,828
Public Administration	0.002	0.049	8,027	0.017	0.130	87,828

Table 2 (continued)
Descriptive Statistics By Presence of Employee Ownership

	Share of firm owned by employees >= 5%			Share of firm owned by employees < 5%		
	Mean	Std. Dev.	Obs	Mean	Std. Dev.	Obs
Firm Death	0.049	0.216	3,342	0.072	0.259	92,513
Strict Firm Death	0.002	0.042	3,342	0.004	0.060	92,513
Employees (in 1000's)	12.402	38.493	3,254	5.809	32.583	79,646
Collective Bargaining	0.254	0.435	3,342	0.071	0.257	92,513
Firm reported any EO stock	1	0	3,342	0.158	0.365	92,513
Firm reported any ESOPs	0.737	0.440	3,342	0.060	0.238	92,513
Value of EO stock per worker	29.915	42.973	3,249	1.266	7.382	92,164
Share of firm owned by employees	0.111	0.091	3,342	0.003	0.008	67,862
Share of firm owned by employees >= 5%	1	0	3,342	0	0	92,513
Share of firm owned by employees >= 10%	0.367	0.482	3,342	0	0	92,513
Agriculture	0.004	0.062	3,342	0.003	0.055	92,513
Mining	0.004	0.062	3,342	0.046	0.210	92,513
Construction	0.009	0.096	3,342	0.009	0.092	92,513
Manufacturing	0.375	0.484	3,342	0.356	0.479	92,513
Transportation	0.116	0.321	3,342	0.094	0.291	92,513
Wholesale Trade	0.033	0.178	3,342	0.030	0.170	92,513
Retail Trade	0.049	0.215	3,342	0.051	0.220	92,513
Finance	0.330	0.470	3,342	0.214	0.410	92,513
Services	0.074	0.262	3,342	0.181	0.385	92,513
Public Administration	0.006	0.075	3,342	0.016	0.126	92,513

Table 3
Relationship Between Employee Ownership and Firm Survival:
Cox Proportional Hazard Regressions Predicting Firm Survival Over 1999-2010

Panel A: Firm reported any EO stock

	(1) Firm Death	(2) Strict Firm Death
Firm reported any EO Stock	0.786***	0.928
	(0.0291)	(0.164)
Employees (in 1000's)	0.982***	0.925***
	(0.00247)	(0.0238)
Collective Bargaining	0.964	1.056
	(0.0551)	(0.301)
Mining	0.950	0.599
	(0.252)	(0.629)
Construction	1.236	-
	(0.361)	-
Manufacturing	1.184	0.656
	(0.305)	(0.664)
Transportation	1.128	0.540
	(0.293)	(0.559)
Wholesale Trade	1.423	0.933
	(0.378)	(0.985)
Retail Trade	1.618*	1.712
	(0.423)	(1.763)
Finance, Insurance and Real Estate	1.013	0.452
	(0.262)	(0.461)
Services	1.677**	0.979
	(0.432)	(0.994)
Public Administration	1.384	6.430*
	(0.367)	(6.536)
Observations	82,900	82,900
Number of Firms	12461	12461
Number of Failures	6100	303
Time at Risk	1.890e+07	1.890e+07
Log Pseudolikelihood	-54201	-2588

Note: Column entries are hazard ratios from Cox proportional hazard regressions. Standard errors are clustered by firm and are in parentheses.

*** $p<0.01$, ** $p<0.05$, * $p<0.1$

Table 3 (continued)
Relationship Between Employee Ownership and Firm Survival:
Cox Proportional Hazard Regressions Predicting Firm Survival Over 1999-2010

PANEL B: Firm reported any ESOPs

	(1) Firm Death	(2) Strict Firm Death
Firm reported any ESOPs	0.821***	0.900
	(0.0453)	(0.258)
Employees (in 1000's)	0.981***	0.924***
	(0.00257)	(0.0236)
Collective Bargaining	0.935	1.053
	(0.0533)	(0.305)
Mining	0.939	0.595
	(0.246)	(0.625)
Construction	1.212	-
	(0.351)	-
Manufacturing	1.170	0.652
	(0.298)	(0.660)
Transportation	1.120	0.538
	(0.288)	(0.557)
Wholesale Trade	1.410	0.929
	(0.371)	(0.982)
Retail Trade	1.593*	1.702
	(0.412)	(1.753)
Finance, Insurance and Real Estate	1.004	0.451
	(0.256)	(0.460)
Services	1.670**	0.975
	(0.425)	(0.990)
Public Administration	1.395	6.417*
	(0.367)	(6.525)
Observations	82,900	82,900
Number of Firms	12461	12461
Number of Failures	6100	303
Time at Risk	1.890e+07	1.890e+07
Log Pseudolikelihood	-54216	-2588

Note: Column entries are hazard ratios from Cox proportional hazard regressions. Standard errors are clustered by firm and are in parentheses.

*** p<0.01, ** p<0.05, * p<0.1

Table 3 (continued)
Relationship Between Employee Ownership and Firm Survival:
Cox Proportional Hazard Regressions Predicting Firm Survival Over 1999-2010

PANEL C: Value of EO stock per worker (in $1000's)

	(1) Firm Death	(2) Strict Firm Death
Value of EO stock per worker (in %1000's)	0.995**	0.957
	(0.00225)	(0.0505)
Employees (in 1000's)	0.980***	0.928***
	(0.00258)	(0.0222)
Collective Bargaining	0.920	1.132
	(0.0523)	(0.338)
Mining	0.958	0.607
	(0.252)	(0.638)
Construction	1.227	0
	(0.357)	(0)
Manufacturing	1.184	0.644
	(0.303)	(0.652)
Transportation	1.140	0.552
	(0.295)	(0.570)
Wholesale Trade	1.424	0.911
	(0.376)	(0.964)
Retail Trade	1.608*	1.632
	(0.418)	(1.679)
Finance, Insurance and Real Estate	1.010	0.470
	(0.259)	(0.479)
Services	1.691**	0.950
	(0.433)	(0.964)
Public Administration	1.414	6.231*
	(0.373)	(6.333)
Observations	82,894	82,894
Number of Firms	12461	12461
Number of Failures	6100	303
Time at Risk	1.890e+07	1.890e+07
Log Pseudolikelihood	-54216	-2584

Note: Column entries are hazard ratios from Cox proportional hazard regressions. Standard errors are clustered by firm and are in parentheses.

*** p<0.01, ** p<0.05, * p<0.1

Table 3 (continued)
Relationship Between Employee Ownership and Firm Survival:
Cox Proportional Hazard Regressions Predicting Firm Survival Over 1999-2010

PANEL D: Share of firm owned by employees greater than or equal to 5%

	(1) Firm Death	(2) Strict Firm Death
Share of firm owned by employees >= 5%	0.772***	0.813
	(0.0637)	(0.337)
Employees (in 1000's)	0.980***	0.924***
	(0.00257)	(0.0237)
Collective Bargaining	0.920	1.050
	(0.0521)	(0.296)
Mining	0.945	0.594
	(0.248)	(0.624)
Construction	1.222	0
	(0.355)	(0)
Manufacturing	1.184	0.653
	(0.302)	(0.661)
Transportation	1.128	0.538
	(0.291)	(0.556)
Wholesale Trade	1.421	0.930
	(0.375)	(0.982)
Retail Trade	1.612*	1.703
	(0.418)	(1.753)
Finance, Insurance and Real Estate	1.002	0.450
	(0.257)	(0.459)
Services	1.688**	0.976
	(0.431)	(0.990)
Public Administration	1.414	6.429*
	(0.372)	(6.534)
Observations	82,900	82,900
Number of Firms	12461	12461
Number of Failures	6100	303
Time at Risk	1.890e+07	1.890e+07
Log Pseudolikelihood	-54217	-2588

Note: Column entries are hazard ratios from Cox proportional hazard regressions. Standard errors are clustered by firm and are in parentheses.

*** $p<0.01$, ** $p<0.05$, * $p<0.1$

IV. LERA Competitive Papers

Assessing the Department of Labor's "My Next Move"

DANIEL J. KOYS
DePaul University

Abstract

The Occupational Information Network (O*NET) contains information on 900+ occupations. My Next Move is an O*NET tool to help people learn about those occupations. A person completes an interest inventory and then receives a list of occupations fitting one's interest profile. In my assessment study, undergraduate students completed a pre-test attitude survey about their career management skills. Then they did an assignment based on My Next Move and completed a post-test attitude survey. There were significant increases in perceived career preparedness skills after doing the assignment. These results are applied to occupations relevant to labor and employment relations degree programs.

Assessing Undergraduates' Career Preparedness Skills

The Higher Learning Commission of the North Central Association accredits colleges and universities. Criterion 4B for this accreditation is that the school must assess its degree programs (Higher Learning Commission 2014). My business school had just gone through its AACSB maintenance of accreditation process (AACSB 2014), so it wanted to do a different assessment for the Higher Learning Commission.

The college's and the university's program review committees agreed to review the undergraduate business program with a focus on career preparedness skills. That review showed that the students needed more help in developing those skills. The college's Undergraduate Curriculum Committee directed each department's curriculum committee to develop its own method to improve the career preparedness skills of its undergraduate students.

The Undergraduate Curriculum Committee asked the departmental curriculum committees to address three career preparedness competencies. As stated in directions to the departments, each of the three had to be directed towards specific outcomes:

1. Self-management skills – These relate to awareness and knowledge of self and utilizing the knowledge of personal and professional success.
    a. Outcome – Being aware of one's career-related interests, abilities, values, and goals.
    b. Outcome – Having the ability to identify opportunities that align with personal interests, abilities, values, and goals.
2. Career building skills – Career building skills relate to finding, securing, and maintaining employment.
    a. Outcome – Being familiar with one's industry.
    b. Outcome – Knowing how to apply for employment.
    c. Outcome – Representing one's skills and abilities in a way that is attractive to employers.
    d. Outcome – Creating social capital (networking).
    e. Outcome – Identifying and choosing best opportunities for advancement.
3. General business skills – General business skills are cross-disciplinary skills across all job types.
    a. Outcome – Demonstrating initiative.
    b. Outcome – Demonstrating organizing and planning.
    c. Outcome – Demonstrating technology awareness and utilization.

A team of faculty members in my department set out to help students obtain the above competencies and outcomes. A two-pronged effort was undertaken. One was intended to develop the self-management skills and general business skills of all the college's undergraduates. This would be accomplished by assigning the Department of Labor's My Next Move exercise in one of the department's core courses, Principles of Management. The second prong was intended to address the career-building skills among the department's own majors. This would be accomplished by a new, half-term course called Career Management Skills.

This research paper describes how I assessed the success of the first prong of the career preparedness effort, that is, the My Next Move exercise.

O*NET and the "My Next Move" Exercise

The U.S. Department of Labor sponsors the Occupational Information Network (O*NET 2014a). That network contains data on over 900 occupations in the United States. Among the information included are each occupation's tasks, knowledge and skills required, interests, work values, wages, and employment outlook.

In a comprehensive study about O*NET, the National Research Council (2010) found that the most common uses of the database are in job analysis and job descriptions in support of different HR activities (primarily recruitment and selection). For example, O*NET's Generalized Work Activities are used as job analysis data in an effort to identify potential selection tools for hiring employees (Jeanneret and Strong 2006).

O*NET database is also used to help people explore career options. Librarians use it to help patrons explore career possibilities and vocational readiness (Bird and Williams 2014). Veterans use it to help them find civilian jobs after leaving military service (Davenport, Richardson, Perkins 2012). It is used as a source of occupational information for systematic person-occupation matching (Converse, Oswald, Gillespie, Field, and Bizot 2004; Kasper 2004).

Of particular interest to this paper is that U.S. colleges are using the O*NET database to help their students. For example, one college links college majors in the U.S. to the 27 knowledge content areas in the database (Freeman and Hirsch 2007). At another college, the Society for Human Resource Management's college curriculum guidelines are found to be consistent with the tasks included in the seven professional HR occupations in the O*NET database (Koys 2013). That college's HR concentration was revised based on those results. A third college uses the database to develop a career and education information web site for its students (Robinson, Meyer, Prince, McLean, and Low 2000).

O*NET also includes tools to help managers, employees, and students learn about the 900 occupations included in its database. One of those tools is My Next Move, "a web-based interactive tool for new job seekers, students, and other career explorers to learn more about their career options" (O*NET 2014a: 1).

In My Next Move (O*NET 2014b), a person responds to a 60-item survey about his or her career interests. After completing the survey, the person receives a score on each of Holland's (1997) six dimensions of work interests (Rounds, Armstrong, Liao, Lewis, and Rivkin 2008). Here is how My Next Move describes the six dimensions (O*NET 2014c):

- "Realistic – People with Realistic interests like work that includes practical, hands-on problems and answers. Often people with Realistic interests do not like careers that involve paperwork or working closely with others."
- "Investigative – People with Investigative interests like work that has to do with ideas and thinking rather than physical activity or leading people."
- "Artistic – People with Artistic interests like work that deals with the artistic side of things, such as acting, music, art, and design."
- "Social – People with Social interests like working with others to help them learn and grow. They like working with people more than working with objects, machines, or information."
- "Enterprising – People with Enterprising interests like work that has to do with starting up and carrying out business projects. These people like taking action rather than thinking about things."

- "Conventional – People with conventional interests like work that follows set procedures and routines. They prefer working with information and paying attention to details rather than working with ideas."

Once people complete the 60-item interest inventory, My Next Move asks them to choose a job level that they think is right for them. They can choose either a job zone that is right for them currently, or they can choose a job zone that will be right at some time in the future (for example, after graduation from college). My Next Move has five possible job zones from which to choose. These reflect the amount of preparation required for an occupation, and they were developed by job analysts based on data supplied by job incumbents. The five job zones are:

1. Little or no job preparation
2. Some job preparation
3. Medium job preparation
4. High job preparation
5. Extensive job preparation

Most undergraduate students fit into job zone 3 while they are in college. Once they graduate and get some work experience, most of them will fit into job zone 4.

Once My Next Move knows the person's interests and intended job zone, it supplies the person with a list of occupations that fit the person's interest profile. By clicking on an occupation title, the person sees a description including that occupation's common job titles, tasks, knowledge, skills, abilities, appropriate personality types, technology used, education usually needed, employment outlook, and other related occupations.

Students in my Principles of Management courses were directed to choose one occupation to explore in detail and to write a two-page (single-spaced) report. The report was to cover the following:

1. From the list of O*NET occupations that matched your interests, which occupation did you select? Why?
2. For your chosen occupation in O*NET, which tasks would be most interesting to you? Least interesting?
3. To what extent did the knowledge areas listed in the O*NET description fit with your existing knowledge areas? What implications does that have for your future course selections at DePaul University?
4. What additional training/experience should you get to obtain the skills and abilities listed in the O*NET description? How will you get that training/experience?
5. To what extent did the education level listed in the O*NET description fit with your current or desired level of education?
6. What implications does the national and state job outlook have for your job search activities and career plans?
7. What implications does the national and state salary ranges have for your job search activities and career plans?
8. Overall, what was the most useful information that you received from completing the self-assessment and occupational research?
9. Find one current job opening or internship listing that matches the results of your self-assessment and the occupation you would like to pursue. Do you think you will apply for this job/internship? Why or why not?
10. Attach the printed/saved results that O*NET gave you to the end of your report. Then attach the job opening or internship listing from step 9 above.

I graded the students' 10 answers using a grading rubric. The assignment was worth 20 points (out of a total of 300 points in the entire course).

Measuring Student Attitudes Towards Career Preparedness

My hypothesis was that there would be a significant difference between the students' perceived career preparedness skills before and after doing the My Next Move exercise. A colleague and I designed a 20-item questionnaire to measure attitudes related to the competencies and outcomes prescribed by the college's Undergraduate Curriculum Committee (see above). The first 13 items measured attitudes towards career interests, career-related abilities, career-related values, career goals, and career exploration. All items used a seven-point Likert scale where 1 = strongly disagree and 7 = strongly agree. The other items asked for demographic information such as year of expected graduation, major, gender, and ethnicity.

I asked the students in my Principles of Management course to complete the questionnaire before they did the My Next Move exercise. After they did the exercise, I asked them to complete the same questionnaire again.

Table 1 presents the characteristics of all the students who completed the questionnaire. As seen in that table, not all of the students completed the pretest and the posttest questionnaires. However, the table shows that the characteristics of the two groups were very similar.

Table 2 presents the assessment of the My Next Move exercise. T-tests were calculated to determine if there was a statistically significant difference between the pretest scores and the posttest scores. Of the 13 items, 12 show statistically significant differences in the pretest scores and the posttest scores.

Table 1

Demographics of the Students Who Completed the Survey

Demographic item	Options	Pretest n (%)	Posttest n (%)
Class standing	Freshman	0 (0.0)	0 (0.0)
	Sophomore	12 (6.8)	9 (5.5)
	Junior	148 (84.1)	142 (86.1)
	Senior	16 (9.1)	14 (8.5)
Expected year of graduation	2014	6 (3.4)	6 (3.6)
	2015	64 (36.4)	62 (37.6)
	2016	105 (59.7)	95 (57.6)
	2017	1 (0.6)	2 (1.2)
Declared major	Accountancy	45 (25.6)	41 (24.8)
	Business Adm.	16 (9.1)	18 (10.9)
	Economics	5 (2.8)	6 (3.6)
	Finance	34 (19.3)	33 (20.0)
	Hospitality Leadership	9 (5.1)	9 (5.5)
	Management	19 (10.8)	16 (9.7)
	M.I.S.	2 (1.1)	2 (1.2)
	Marketing	10 (5.7)	7 (4.2)
	Real Estate	2 (1.1)	1 (0.6)
	Other	12 (6.8)	9 (5.5)
Gender	Male	106 (60.2)	97 (58.8)
	Female	70 (39.8)	68 (41.2)
Race/ethnic group	Black	9 (5.1)	9 (5.5)
	American Indian	1 (0.6)	0 (0.0)
	Asian	22 (12.5)	21 (12.7)
	White	126 (71.6)	114 (69.1)
	Hispanic	13 (7.4)	12 (7.3)
	Multi-Racial	1 (0.6)	1 (0.6)
	I'd rather not say	4 (2.3)	8 (4.8)

Table 2
Attitudes Towards Career Skills: Before and After Doing the "My Next Move" Assignment

Item	Pre- or Posttest	n	Mean	s.d.	t	df	Sig. 2-tail
I have a strong awareness of my career interests.	Pretest Posttest	166 157	5.73 6.10	1.21 1.12	-2.77	321	.006
I have a strong awareness of my career-related abilities.	Pretest Posttest	166 157	5.79 6.18	0.90 0.95	-3.78	321	.000
I have a strong awareness of my career-related values.	Pretest Posttest	166 157	5.98 6.23	0.83 0.82	-2.76	321	.006
I have a strong awareness of my career goals.	Pretest Posttest	166 157	5.86 6.07	1.14 1.10	-1.72	321	.086
I can identify career opportunities that align with my career interests.	Pretest Posttest	166 157	5.66 6.11	1.22 0.93	-3.79	321	.000
I can identify career opportunities that align with my career-related abilities.	Pretest Posttest	166 157	5.59 6.18	1.03 0.81	-5.68	321	.000
I can identify career opportunities that align with my career-related values.	Pretest Posttest	166 157	5.57 6.07	1.05 0.92	-4.57	321	.000
I can identify career opportunities that align with my career goals.	Pretest Posttest	166 157	5.70 6.01	0.96 1.02	-2.75	321	.006
I am familiar with my intended occupation (that is, a type of job such as an HR Specialist, an Accountant, or a Market Researcher).	Pretest Posttest	166 157	5.54 6.02	1.37 1.07	-3.47	321	.001
I am familiar with my intended industry (that is, the type of product/service like hospitality, health care, or retail trade).	Pretest Posttest	166 157	5.65 6.12	1.22 0.98	-3.80	321	.000
I have displayed initiative (for example, seeking out internships, developing a personal network, completing occupational inventories, researching possible careers) regarding my personal career management.	Pretest Posttest	166 157	5.45 6.04	1.53 1.26	-3.79	321	.000
I have the organizing and planning skills (for example, courses mapped out to develop needed career knowledge, understanding the national and local job outlook, understanding pay ranges) to help with my personal career.	Pretest Posttest	166 157	5.39 6.11	1.33 1.03	-5.43	321	.000
I know how to use technology (for example, O*NET, LinkedIn) to help me understand and manage my career path.	Pretest Posttest	166 157	5.77 6.51	1.20 0.69	-6.70	321	.000

Results Show That "My Next Move" Increased Career Preparedness Skills

The statistical analysis supported my hypothesis that there would be a significant difference between the students' perceived career preparedness skills before and after doing the My Next Move exercise. However, a statistically significant difference is not always practically significant. The assignment produced practically significant differences on eight of the 13 items.

The biggest practical differences were on the following four items. They showed a difference of over one-half a point on the seven-point scale (like moving from a mean score of 5.5 to a mean score above 6.0). These items are mainly related to the competency called general business skills.

- I know how to use technology (for example O*NET, LinkedIn) to help me understand and manage my career path. [There was an average increase of 0.74.]
- I have the organizing and planning skills (for example, courses mapped out to develop needed career knowledge, understanding the national and local job outlook, understanding pay ranges) regarding my personal career management. [There was an average increase of 0.72.]
- I have displayed initiative (for example, seeking out internships, developing a personal network, completing occupational inventories, researching possible careers) regarding my personal career management. [There was an increase of 0.59.]
- I can identify career opportunities that align with my career-related abilities. [There was an increase of 0.59.]

Four additional items showed changes large enough to be considered practical differences, but their mean scores only moved about one-half a point on the seven-point scale. These items are mainly related to identifying career-related information.

- I can identify career opportunities that align with my career-related values. [There was an average increase of 0.50.]
- I am familiar with my intended occupation (that is, a type of job such as HR Specialist, an Accountant, or a Market Researcher). [There was an average increase of 0.48.]
- I am familiar with my intended industry (that is, they type of product/service list hospitality, health care, or retail trade). [There was an average increase of 0.47.]
- I can identify career opportunities that align with my career interests. [There was an average increase of 0.45.]

The My Next Move assignment did not produce practically significant changes in the perceived career preparedness skills for five items. Their mean scores only moved about one-third of a point on the seven-point scale. These items are primarily related to self-awareness.

- I have a strong awareness of my career-related abilities. [There was an average increase of 0.39.]
- I have a strong awareness of my career interests. [There was an average increase of 0.37.]
- I can identify career opportunities that align with my career goals. [There was an average increase of 0.31.]
- I have a strong awareness of my career-related values. [There was an average increase of 0.25.]
- I have a strong awareness of my career goals. [There was an average increase of 0.21.]

My assessment of the My Next Move exercise shows that the exercise is successful in helping improve many career preparedness competencies, but not all of them. The exercise was not a big help in regards to improving career-related self-awareness. It was a big help in developing general business skills (showing initiative, organizing and planning, and technology utilization), and it was a moderate help in obtaining information about chosen careers.

Implications for Labor and Employment Relations Degree Programs

Professors teaching in labor and employment relations programs can use My Next Move to help their students develop career preparedness skills and to help them obtain information about labor and employment relations occupations. When one searches O*NET's database for "labor and employment relations" occupations, the first ten occupations that O*NET returns are:

- Labor relations specialists[1]
- Human resource managers
- Human resource specialists
- First-line supervisors of helpers and laborers
- Training and development specialists
- Compensation and benefits managers
- Training and development managers
- Compensation, benefits, and job analysis specialists
- Business operations specialists, other
- Arbitrators, mediators, and conciliators

Almost all of those occupations are high on Holland's career interests of Enterprising, Conventional, and Social (see Table 3). As defined above, people with Enterprising interests like work that has to do with starting up and carrying out business projects; these people like taking action rather than thinking about things. People with Conventional interests like work that follows a set of procedures and routines; they prefer working with information and paying attention to details rather than working with ideas. People with Social interests like working with others to help them learn and grow; they like working with people more than working with objects, machines, or information.

Table 3
Labor and Employment Relations Occupations and Holland's Career Interests

	Mean Job Incumbent's Career Interest Score in O*NET		
O*NET Occupations	**Enterprising**	**Conventional**	**Social**
Labor relations specialists	89	61	56
Human resource managers	100	67	78
Human resource specialists	83	61	50
First-line supervisors of helpers and laborers	100	56	
Training & development specialists	50	61	95
Compensation & benefits managers	100	72	61
Training & development managers	100	50	83
Compensation, benefits, and job analysis specialists	89	89	
Business operations specialists, other	n/a	n/a	n/a
Arbitrators, mediators, and conciliators	89	50	95

If professors in labor and employment relations degree programs use My Next Move in their courses, they are likely to find that their students have the above interests. They could then teach their courses with those types of interests in mind. For example, in addition to covering economic, legal, and psychological theories, their courses could include business projects that are action-oriented rather than idea-oriented. Their courses could also include details about policies and procedures related to staffing, training, pay & benefits, and collectively bargained contracts. Finally, their courses could address social topics, such as how to be of service to other people and how to help other people learn and grow.

Acknowledgments

Jaclyn Jensen assisted in the development of the 20-item questionnaire that measured attitudes related to career management competencies and outcomes.

Endnote

[1] According to Phil Lewis of the National Center for O*NET Development, O*NET's description of the Labor Relations Specialist has recently been separated from that of the Human Resource Specialist. A further update will be published in July 2015 in the O*NET 20.0 developmental database. It will then be incorporated within the My Next Move sites and O*NET online two to three months later.

References

AACSB International. 2014. "Accreditation Standards." http://www.aacsb.edu/en/accreditation/standards/. [December 15, 2015].

Bird, Nora J. and Ted Williams. 2014 "Casting a Wider Net: O*NET, Workforce Development, and Information Literacy." *RUSQ: Reference and User Service Quarterly, Vol. 23,* no. 3, pp. 227-231.

Converse, Patrick D., Frederick L. Oswald, Michael A. Gillespie, Kevin A. Field, and Elizabeth B. Bizot. 2004. "Matching Individuals to Occupations Using Abilities and the O*NET: Issues and an Application in Career Guidance." *Personnel Psychology, Vol. 57,* pp. 451-487.

Davenport, Katie, Cameron Richardson, and Daniel Perkins. 2012. "An Initial Review of Employment Resources." Unpublished paper, The Clearinghouse for Military Family Readiness, The Penn State University.

Freeman, James A. and Barry T. Hirsch. 2007. "College Majors and the Knowledge Content of Jobs." Unpublished paper, Institute for the Study of Labor (IZA), Bonn, Germany.

Higher Learning Commission. 2014. "The Criteria for Accreditation and Core Components." https://www.ncahlc.org/Criteria-Eligibility-and-Candidacy/criteria-and-core-components.html. [December 15, 2015].

Holland, John L. 1997. *Making Vocational Choices: A Theory of Vocational Personalities and Work Environments* (third edition). Odessa, Florida: Psychological Assessment Resources.

Jeanneret, P. Richard and Mark H. Strong. 2006. "Linking O*NET Job Analysis Information to Job Requirement Predictors: An O*NET Application." *Personnel Psychology, Vol. 56,* no. 2, pp. 465-492.

Kasper, Henry T. 2004. "Matching Yourself with the World of Work: 2004." *Occupational Outlook Quarterly, (Fall),* pp. 2-21.

Koys, Daniel. 2013. "Using the Department of Labor's O*NET Database in Teaching HRM." *Journal of Human Resources Education, Vol. 7,* nos. 3/4, pp. 1-21.

National Research Council. 2010. *A Database for a Changing Economy: Review of the Occupational Information Network (O*NET).* Panel to review the Occupational Information Network (O*NET). Nancy T. Tippins & Margaret L. Hilton, editors. Committee on National Statistics, Division of Behavioral and Social Sciences and Education. Washington, D.C.: The National Academy Press.

O*NET. 2014a. "O*NET Resource Center." http://www.onetcenter.org/mynextmove.html. [December 15, 2014].

O*NET. 2014b. "My Next Move" http://www.mynextmove.org/. [December 15, 2014].

O*NET. 2014c. "O*NET Interest Profiler." http://www.mynextmove.org/explore/ip. [December 15, 2014].

Robinson, Nicholas K., Dinorah Meyer, Jeffery P. Prince, Christopher McLean, and Richard Low. 2000. "Mining the Internet for Career Information: A Model Approach for College Students." *Journal of Career Assessment, Vol. 8,* no. 1, pp. 37-54.

Rounds, James, Patrick I. Armstrong, Hsin-Ya Liao, Phil Lewis, and David Rivkin. 2008. "Second Generation Occupational Interest Profiles for the O*NET System: Summary." Unpublished paper, National Center for O*NET Development, Raleigh, NC.

V. New Perspectives on Occupational Closure: Licensure, Certification, and Unionization

Analyzing the Influence of Occupational Licensing Duration on the Labor Market

SUYOUN HAN
MORRIS M. KLEINER
University of Minnesota

Abstract

We analyze the economic impact of the duration of occupational licensing statutes for twelve major licensed occupations in the U.S. labor market. Time from the start of occupational licensing (i.e. licensing duration) matters in influencing the economic outcomes because states often enact grandfather clauses that protect existing workers, increasing the cost of new entrants. Our non-parametric Kernel estimation techniques suggest that it takes at least 10 years for the labor market to absorb and capitalize the full effects of occupational licensing.

Introduction

Occupational licensure, the legal process by which governments establish qualifications to practice a trade or profession for pay, has become one of the fastest growing and significant labor market regulations in the United States. For example, the proportion of all workers covered by occupational regulations increased from about 5 percent in the early 1950s to almost 29 percent in 2008, which is substantially greater than the portion of all workers covered by either minimum wage legislation or unionization (Kleiner, 2000; Kleiner and Krueger, 2010; Kleiner and Krueger, 2013). The Council of State Governments estimated that about 800 occupations were licensed by at least one state in the 1990s (Brinegar and Schmitt, 1992). Given the pervasiveness and growth of occupational licensing, it has become a key issue in the access of workers to jobs and their potential labor market and economic outcomes.

Recent analysis of occupational licensing has shown the influence of licensing on wage determination (Kleiner, 2000; Kleiner and Kudrle, 2000; Kleiner and Todd, 2009, Kleiner and Krueger, 2010). Estimates developed by Kleiner and Krueger (2013) show that occupational licensing is associated with 15% higher wages, which is consistent with the economic theory that occupational licensing restricts the supply of labor to the occupation resulting in an increase in wages (Rottenberg, 1980).

There has been little research devoted to understanding how licensing duration affects economic outcomes, however one exception had been the research of Thornton and Timmons (2013). They examine the relationship between licensing duration and economic outcomes for massage therapists, finding that duration of licensing by state is an important factor in wage determination. However, their OLS estimates are confounded by selection bias as licensing duration could be longer for highly skilled members of an occupation that are more likely to receive higher wages.

Another recent study has examined the influence of the duration of licensing on wage determination for nurses and found some evidence that wages of nurses increased with the duration of licensing, but they found no effects on employment in the occupation (Law and Marks, 2014). Other studies for the UK find that the length of licensing duration has a positive influence on wage determination for a large number of licensed occupations (Bryson, et al. 2012).

Grandfathering and Licensing Duration

The time from the passage of occupational licensing laws may be important in analyzing occupational licensing. One rationale is that states often enact "grandfather clauses" that protect existing workers by allowing them to practice even though they may not meet the current requirements, and new entrants must have higher entry standards than existing members of the occupation. The process of older lesser educated workers leaving, and newer workers with higher entry requirements coming into the occupation takes time to work its way through the labor market. This process may limit the supply of labor and allow those in the occupation to gain economic benefits by limiting employment growth. Therefore, licensing duration—the time from the implementation of occupational licensing legislation—matters because it may take several years for the full effects of occupational licensing to be realized in the labor market.

Statutory Information and Labor Market Data

Measures of Licensing Duration

We collected statutory data on licensing duration in the labor market for major twelve licensed occupations from the various data sources. First, the Council of State Governments has collected information on the year each state passed their first licensing legislation for several licensed occupations. Second, we examined the Lexis Nexis law databases and obtained the remaining information including major changes in the licensing statutes and administrative procedures. Third, we contacted the Supreme Court library and Board of Examiners for the states where we were not able to get the precise year of the legislation from the Council of State Governments or from Lexis Nexis. The basic data for the duration of each of the occupations in our analysis are shown in Figure 1.

Figure 1
Timeline Showing When Each of the Occupations Became Licensed by State

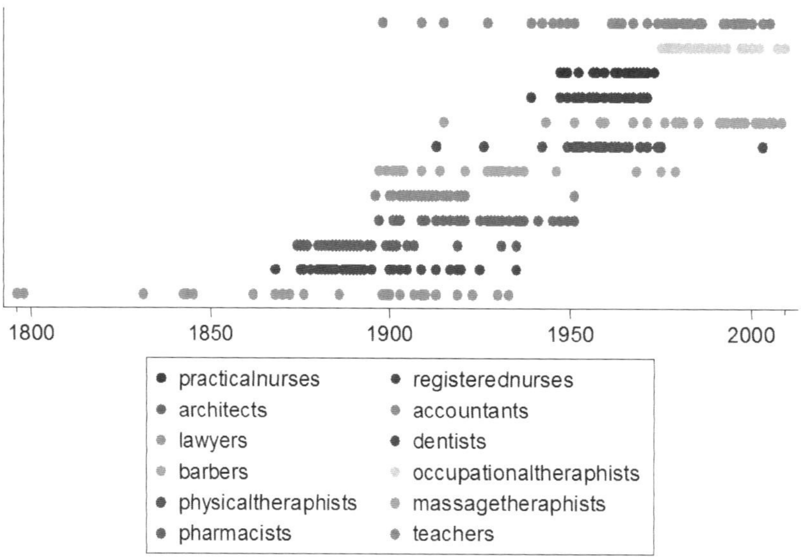

In 2013, using estimates from the American Community Survey (ACS), 11 million persons worked in these twelve licensed occupations. Moreover, these twelve occupations cover a range of occupations from blue to white collar and from old to young in terms of licensing duration. For each state and occupation we develop a duration value by calculating current year minus year when each state first implemented its first licensure law.

Wages and Labor Supply

In order to develop a sufficient time line we use all available data from the Census and ACS for years from 1940 to 2013. We restrict the sample to individuals who worked in major twelve licensed occupations or individuals who worked in unlicensed occupations. We construct our sample from 1940 to 2013 each of the occupations and were able to check whether the occupation is licensed or unlicensed (Meyer and Osborne, 2005). Finally, our sample includes individuals who were either in the major universally licensed occupations or unlicensed occupations. We define individuals who worked in major licensed occupations as a treatment group, and individuals who worked in all other unlicensed occupations as a comparison or control group.

Next, we include standard human capital variables from the ACS such as gender, age, education, and potential experience. In order to generate a reliable sample for our analysis, individuals whose education are "below 12th grade without a diploma" for dentists, lawyers, accountants, and pharmacists were dropped. In addition, individuals whose age is greater than 65 and whose years of potential experience are estimated to be below zero also were deleted.

Finally, hourly real earnings were determined by dividing the annual earnings by annual hours worked, measured in the 1940 Consumer Price Index (CPI). Annual hours worked were calculated by multiplying the usual working hours to the number of weeks for the past 12 months. We eliminated from the sample those with more than 60 hours of work in a week. In addition to all these restrictions, the original sample is trim down by excluding individuals with wages below the federal minimum wage level. The resulting sample consists of 15,120,377 observations from 1940 to 2013 using the Census and ACS.

Model and Results

Nonparametric Kernel Estimation

For the model of the effects of the occupational licensing duration on wage determination we use a nonparametric kernel estimation procedure shown in Figure 2. This process is a form of data smoothing. We try to visually inspect whether the longer duration of licensing exhibits greater earnings in distribution. The figure in the solid line represent the wage distribution for the unlicensed workers, the dashed one is the distribution for the licensed workers whose licensing duration are less than 10 years and the dashed and doted one is the distribution for the licensed workers whose licensing durations are more than 10 years. Our non-parametric kernel estimation results suggest that it takes at least 10 years to fully realize the full economic effects of occupational licensing.

Figure 2
Distribution of Log Hourly Earnings by Occupational Licensing Duration

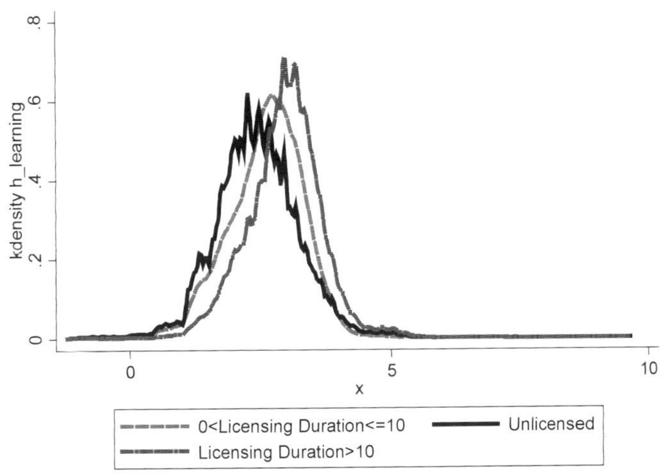

Census 1940-2000 and American Community Survey (ACS) 2001-2013

Conclusions

Since occupational licensing takes time to fully implement, duration should matter in influencing labor market outcomes. For example, states often enact "grandfather clauses" that protect existing workers, and new entrants must have higher entry standards than those already in the occupation. The process of older lesser educated workers leaving, and newer workers with higher entry requirements coming into the occupation takes time to work its way through the labor market. The results of our study should allow policy analysts and makers to develop and implement more informed decisions on this rapidly growing labor market institution.

References

Brinegar, Pamela L. and Kara L. Schmitt. 1992. "State Occupational and Professional Licensure." *The Book of the States*, 1992-1993. Lexington KY: The Council of State Governments, pp. 567-580.

Bryson, Alex, John Forth, Amy Humphris, Morris M. Kleiner, and Maria Koumenta. 2012. "The Incidence and Labor Market Outcomes of Occupational Regulation in the UK." Paper presented at the Allied Social Science Associations annual meeting, Janua5–8, Chicago.

Council of State Governments .1952. *Occupational Licensing Legislation in the States. Chicago: Council of State Governments.*

Kleiner, Morris M. 2000. "Occupational Licensing," *Journal of Economic Perspectives*, 14(4): 189–202.

Kleiner, Morris M., and Robert T. Kudrle. 2000. "Does Regulation Affect Economic Outcomes? The Case of Dentistry." *Journal of Law and Economics* 43(2): 547–582.

Kleiner, Morris M., and Richard M. Todd. 2009. "Mortgage Broker Regulations that Matter: Analyzing Earnings, Employment, and Outcomes for Consumers." In *Studies of Labor Market Intermediation*, ed. David Autor. Chicago: University of Chicago Press.

Kleiner, Morris M., and Alan B. Krueger. 2010. "The Prevalence and Effects of Occupational Licensing." *British Journal of Industrial Relations* 48 (4): 676–87.

Kleiner, Morris M., and Alan B. Krueger. 2013. "Analyzing the Extent and Influence of Occupational Licensing on the Labor Market." *Journal of Labor Economics* 31(2, Part 2): S173-S202.

Marc, Law T. and Mindy Marks, 2014. "What Are The Long Run Effects of Nurse Occupational Licensure?" Working Paper, University of Vermont. October.

Meyer, P. B., and A. M. Osborne.2005. "Proposed Category System for 1960-2000 Census Occupations." Bureau of Labor Statistics Working paper 383.

Rottenberg, Simon.1980. "Introduction," In *Occupational Licensure and Regulation*, S. Rottenberg, ed. Washington, American Enterprise Institute, pp.1-10.

Thornton, Robert J. and Timmons, J. 2013. "Licensing One of the World's Oldest Professions: Massage." *Journal of Law and Economics* 56(2): 371-388.

VI. LERA Papers I: New Models

Democracy, Militancy, and Union Revitalization: The DeMReV Model of Union Renewal—A Strategic Model Expanding on Voss and Sherman

JERRY A. CARBO
M. BLAKE HARGROVE
STEVEN J. HAASE
Shippensburg University

Abstract

Voss and Sherman (2000) presented a model of union revitalization, based on unions' needs for reform at that time. Since then, the situation has gotten much worse for labor, especially in the United States. Here, we present a new model, drawing on and extending Voss and Sherman (2000). Our extension goes beyond their factors of union reform which are mainly via outside organizers and internal leaders responding to a crisis. Specifically, we illustrate the important role for *rank-and-file* union militancy and democracy for bringing about revitalization. We draw on a number of historical and recent examples of revitalization as well as illustrate additional reactions from the unions themselves (i.e., the *oligarchic response*) to bottom-up democratic revitalization efforts. Our model will hopefully be seen as means to reform for union members and progressive leaders who realize that the status quo is not sustainable, and new tactics are needed in times of crisis like these.

Introduction

Labor unions and the labor movement in general deserve a great deal of credit. In the past century, society as a whole and workers in particular have benefitted from union efforts to create social justice. In the United States, Europe, and to varying degrees throughout the world the labor movement has been the catalyst for positive change. The labor movement drove many of the reforms that Americans take for granted today—the 5 day work week, the 8 hour day, the elimination of child labor, the birth and expansion of the American middle class, the passage of the US Civil Rights Act and other employment legislation; all of these gains were either initiated by or dependent upon the support of strong labor unions. However, with the decline of union density and thus the power of the labor movement, there is growing concern and evidence that these gains may be lost. This decline has been linked to the growing (and record level of) inequality in the United States between the richest and poorest of Americans as well as the losses of prior gains for minority groups in the United States (Rosenfeld 2014). We also see working conditions, from real wages to the lack of benefits to workplace safety, in a decline that coincides with the loss of labor side power. The reasons for this loss of power are many and varied. Many if not most of these reasons come from outside of the labor movement itself (weak laws, weak enforcement, growth of a union busting industry, a corporate/capital friendly political environment, and other factors). However, there are also internal reasons for this decline, including leader and member apathy, and a tendency towards oligarchy within union leadership.

In theory, labor unions are the most democratic institutions in the United States. At its heart, a labor union is nothing more and nothing less than a group of workers engaging in concerted activity for mutual aid and protection—each of these workers with an equal voice, standing in the collective. It is through this collective voice and power that labor unions have been able to secure the myriad of gains for working people in the United States and across the globe in the more than two centuries of labor history. However, in practice, unions often stray from this ideal of democracy and fall prey to a pattern of oligarchical practices

(Voss and Sherman 2000). When this occurs, unions no longer serve members, but rather serve the interests of a few select union leaders. Unions lose the ability or desire to address member concerns and members often become disengaged and apathetic (Moody 2007). With this lost desire and lost ability, prior gains and future gains are also lost. However, in some cases, rank-and-file members *do* respond to this oligarchy through revitalization and reform efforts.

In this paper, we contribute to the literature by providing a path forward for labor leaders to revitalize their unions and the labor movement. The path we present focuses on the strategies and tactics that have proven successful to revitalizing and reforming labor unions during the past fifty years. While there are many case studies of union revitalization that address the strategies and tactics we present in our model, we have been able to collect and to categorize these strategies and tactics into democratic and militant strategies. Further, in this manuscript we will present evidence that union democracy and militancy are synergistic and together present a strong path towards revitalizing any union and the labor movement as a whole. While this is not the only potential path for revitalization, it is a promising path that has proven successful in many eras of the labor movement and a path that deserves more attention and future research.

We first illustrate the context that led to the development of the Voss and Sherman (2000) model and then develop the rationale for expanding their model. In doing so, we illustrate key "tipping points" that motivate union revitalization, with a special emphasis placed on the importance of strong democratic structures within the unions themselves, as well as the importance of utilizing militant tactics for accomplishing goals. Within this context, the role of union leadership is discussed in terms of opposition to reform (e.g., oligarchic vs. democratic response). We then provide case studies of successful outcomes in similar, past union reform efforts. Thus, the narrative of our paper leads us to propose an alternative model, "DeMReV", which accounts for the success of the case studies via Democracy and Militancy leading to Revitalization.

The Voss and Sherman Model

In 2000, Voss and Sherman recognized that the labor movement was in desperate need of reform due to the crisis of dwindling density and influence of unions. They also recognized that at least part of this crisis was brought on by issues internal to the labor movement and unions. Many unions were becoming "more like an institutionalized interest group than a social movement" (p. 304). Unions as democratic institutions are as likely as any other to become oligarchic (Michels 2009). In line with Michels' Iron Law of Oligarchy, many unions were also becoming oligarchical and stagnant. Union leaders look towards tactics and methods of preserving their positions of authority rather than looking out for the interests of the organization and its members. Many unions at the time that Voss and Sherman developed their model were indeed being trapped by this Iron Law of Oligarchy. Voss and Sherman explored whether unions that fell prey to the Iron Law of Oligarchy would thus be able to be reformed and revitalized, and if so, how these reforms could take shape.

At the time Voss and Sherman developed their model of union revitalization, labor was indeed going through a crisis, the same crisis that continues today—declining density and declining rights for working people, coupled with stagnating wages and growing inequality. However, there were reasons for renewed hope in the labor movement. John Sweeney, former President of the SEIU, had recently been ushered in as the President of the AFL-CIO on a reform ticket via a reform movement (Milkman and Voss 2004). Sweeney committed the AFL-CIO to a renewed focus on organizing the unorganized and asked member unions to do the same. The reform efforts were adopted by relatively few member unions of the AFL-CIO, but among those who did adopt such efforts, Sweeney's own former union, the SEIU was considered a leader. As Milkman and Voss (2004) have suggested, many of the unions that did engage in meaningful reform and meaningful organizing efforts did so through plans "orchestrated from the top" often by union staffers who did not adopt the oftentimes risk-averse behavior of elected union leaders. Again, the SEIU, in particular under the guidance of Stephen Lerner, led the way in these efforts, often calling for an autocratic style of "purging any union staffers who refuse to go along with the program" (Milkman and Voss 2004:7). The SEIU was also coming off of their recent successes in organizing service workers via a top down "Justice for Janitors" campaign. Finally, the AFL-CIO Union Summer initiative had helped to develop labor leaders as well as leaders of other social movements who would have an affinity towards the labor movement plight. The environment created a strong reason to support the idea that reform efforts should be top down, especially in the organizing phase. The most successful reform oriented union, the SEIU, had clearly adopted such a style—the former President of the SEIU had been elected to be the President of the AFL-CIO, and the AFL-CIO had adopted programs to develop staffers as labor leaders.

Within this context and after engaging in in-depth qualitative research with UFCW, SEIU and HERE locals, Voss and Sherman answered the possibility of reform in the affirmative and, in answering this question, suggested a model of union reform. The Voss and Sherman model contains three pieces—a crisis, reform efforts, and organizing revitalization. In this model, the effort directed at revitalization is augmented and driven by outside organizers and union leadership at the international level. These efforts are hindered by staffers and even members who are wedded to a more servicing orientation of unionism or opposed to the organizing model. We depict our interpretation of this model in Figure 1. The model also suggests a strong role for education of members at least in part so that they handle the internal problems on their own while staffers focus on organizing. The end result of this revitalization is a union that is more able to organize the unorganized.

Figure 1
Carbo, Haase, & Hargrove's Interpretation of Voss & Sherman's (2000) Model for Union Revitalization

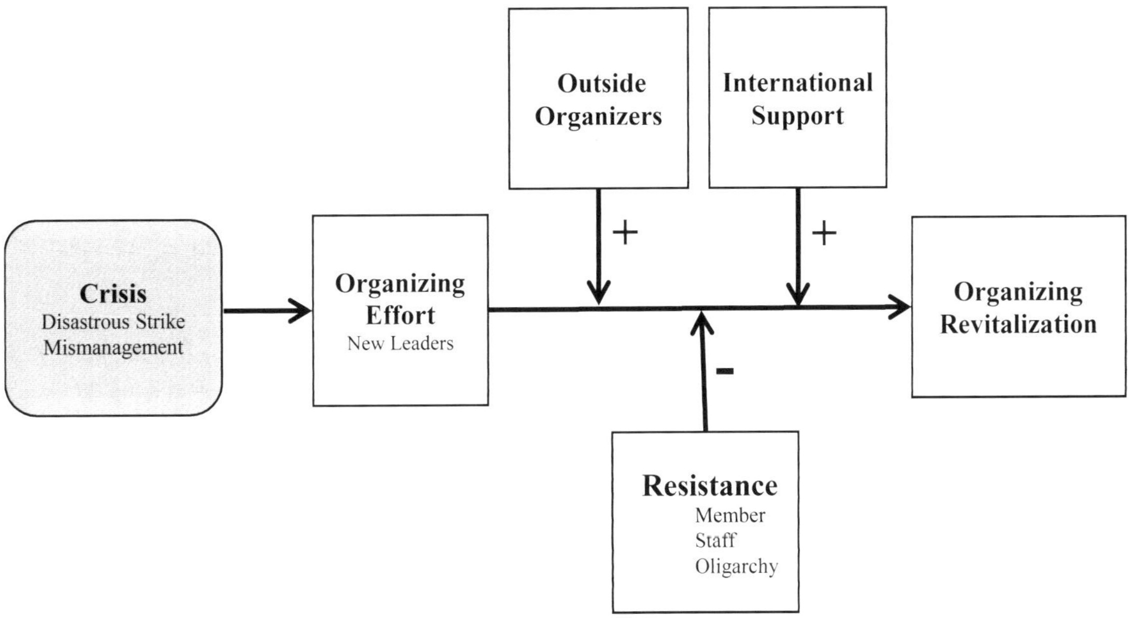

The Voss and Sherman model supports a path forward that requires aggressive union tactics and there continues to be support for the need for these types of tactics. Further, staff support and experience in engaging in these tactics would of course logically make sense as a benefit to such union action. Even the successful reform efforts of HERE under Wilhelm and Sirabella had some components of the top-down approach as an outsider-insider joint effort was adopted with outsiders often bringing in the expertise in organizing, mobilizing and campaigning strategies (Voss and Sherman 2000; Getman 2010). There is little doubt that there is value in this top level or staffer expertise.

However, despite these top-down efforts, the crisis of the labor movement remains the same today as it was in 2000. Much like the contextual influence on the Voss and Sherman model, our research is also influenced by context. Similar to the time of the Voss and Sherman research, labor is marked by decreasing union density and decreasing influence. However, the hopes and the environment of union reform have changed. The corporate style of union leadership of the SEIU has been brought into question by labor academics (Moody 2007) and by their own members' rank-and-file reform effort (SMART). We have seen accusations that SEIU's organizing success has as much to do with poaching already organized members as it does to do with organizing the unorganized. Today, we see different rays of hope in the labor movement, and rank-and-file, bottom-up movements have become forces for positive change. The Occupy movement, the Arab Spring, the Fight For $15, the battles of low-wage workers against corporate giants such as Wal-Mart and McDonalds, have taken the lead in the battle for justice and equity. All of these movements are driven by citizens and workers themselves without centralized leadership or organizational governance. We also see militant and rank-and-file union successes such as the Chicago Teacher's strike, the Chicago Glass Workers, and the successes of the SMART reforms in the SEIU. In light of, and at least in part because of, the evolving context we present a different model for labor reform and revitalization. While Voss and Sherman do indeed mention the need for rank-and-file action to address day to day issues, it is not a focus of their research. We present a strategic model that focuses on these actions of internal reformers, rank-and-file reformers and their utilization of democratic and militant strategies. We also focus on what often becomes an internal struggle between the reformers and the entrenched oligarchy for the hearts and minds of the members.

Defining Revitalization

In order to contextualize our research, it is critical to understand our definition of revitalization and how this may differ from the Voss and Sherman definition as well as other models of revitalization. Generally, revitalization is defined as "giving new life or vigor to" (The Merriam-Webster Dictionary 1974). This general definition fits both our model and the Voss and Sherman (2000) model. Voss and Sherman suggest that revitalization entails "break [ing] out of this bureaucratic conservatism" which has led many unions to become more like "an institutionalized interest group than a social movement" (2000:304). They suggest that revitalization entails the use of disruptive tactics that would be present in social movements and "antithetical" to bureaucratic organizations. In essence, we see that revitalization is a return to a social movement. We agree that revitalization encompasses this concept of once again driving toward the direction of a social movement and breaking free from the binds of bureaucracy. However, the Voss and Sherman model focuses almost exclusively on the growth of the movement (i.e., organizing the unorganized or expanding union density). We believe that revitalization entails more—that it is about a renewed ability to meet the larger goals of the movement.

While Voss and Sherman focus on restoring life to the labor movement through its ability to organize the unorganized, we suggest that by stepping back and looking at the purpose of unions and the labor movement a more expansive definition of revitalization is appropriate. Quite clearly the goal of the labor movement is to improve the lives of working people (Brofenbrenner 2005; Yates 2009). Each union looks to improve the lives of their members and looks to organize other members to improve their lives as well. While union density (organizing success) of course plays a huge role in the process, so does the management of members' day-to-day working lives. Bureaucratic unions not only find that they are ineffective at organizing new members, but they also are ineffective at fighting for member rights during collective bargaining and during the day-to-day management of the CBA (Moody 2007). Therefore, if unions are to meet their ultimate goals, then they must reform not only their organizing techniques, tactics, and strategies, but they must also reform and revitalize the way that they operate on a day-to-day basis and at the bargaining table. We suggest that revitalization should encompass all phases of the labor relations process from the organizing phase to collective bargaining. In other words, we believe revitalization is critical to the day-to-day representation of members. Our definition of reform and revitalization is more in line with Nyden's definition of rank-and-file insurgency which entails "molding the union into a more responsive organization which will protect workers in ALL aspects of their relationship with management" (1984: 2). This definition includes wages and benefits, workers' control over their jobs, and a push for union structures to become more responsive (Nyden 1984). Not only is revitalization about renewed outward vigor (organizing the unorganized),it is also about renewed inward vigor (fighting for gains for members). From this understanding of our definition of revitalization, we can now explore the components of a model that would lead to this form of revitalization.

A Strategic Model of Union Reform and Revitalization

In addition to extending the concept of revitalization from an outward perspective to a combined internal and external perspective, we propose to expand the Voss and Sherman model in a number of other ways. While much of the Voss and Sherman model focused on the actors, our model will focus on the internal actions, strategies, and tactics that are needed to achieve revitalization. Thus, we present our model as a strategic extension of the Voss and Sherman Model of Union Reform and Revitalization. In the following sections, we will define this strategic model of union reform and revitalization from beginning to end. First, we describe the tipping points that lead to union reform efforts. From tipping points, we present evidence that successful reform needs to entail both democratic and militant strategies. We will also address the role of entrenched union leaders in these efforts and will paint a much different picture of leader responses than the one depicted by Voss and Sherman (2000) and Fiorito (2004). Our model will focus on internal and external revitalization efforts, focusing on union success in not only organizing the unorganized but also in achieving gains for members in each phase of the labor relations process.

While the general theories of our research were developed by the context of the current labor movement, including our own union, we first also looked to social science and labor studies research to help us in developing our hypothetical model. We then tested this model by exploring cases studies via a literature review of union reform efforts and examples of successful union strategies and tactics. We first looked for common strategic themes within the reform movements. Where we found common themes, we also took the next step to define and describe the tactics within these strategies. The reform efforts we reviewed ranged from recent efforts to efforts from the middle of the last century. We believe that each of these reform efforts hold lessons for labor researchers, leaders and union members. These reform efforts include the Painters union (NYPPU) reform of the 1960s**,** Fight Back Movement within the Steelworkers, Mineworkers for Democracy, the Teamsters for a Democratic Union, SMART, RMT in Great Britain, the CORE reformers of the Chicago Teacher's Union, and Sirabella's UNITE HERE. We also used Lipset, Trow, and Coleman's (1956) description of the ITU as a guide to the components of union democracy.

Development of the Theoretical model through Contextual Cues and Case Studies

Contextual Cues

Much like the Voss and Sherman model was influenced by the successes of the SEIU and Sweeney models of revitalization, our model has also been influenced by the context of the labor movement and recent events. While there have been clear models of success in terms of organizing under the SEIU, the labor movement continues to see decreases in union density. However, the glimmers of hope today are different than those from the late 1990s and early 2000s. As mentioned earlier, today we see successes of grassroots efforts with strong direct involvement from working people and union members. From the takeover of the Capital building in Wisconsin and the Occupy Movement, to the Fight for $15 and the struggle against police violence, grassroots, militant, democratic efforts have led to successes and to moments of support for the labor movement. These successful movements played a large role in our decision to explore whether these types of militant or aggressive tactics, combined with the type of direct democracy in these movements, are keys to union reform and revitalization.

The Impetus for Change

Any revitalization effort requires that the members of the organization desire a change and that they decide to take actions to bring about that change. Voss and Sherman suggest that it is a crisis that leads to these change efforts. While we believe that a crisis can indeed lead to these change efforts, we also believe that there are times when something less than a crisis will lead to these efforts. According to Chong (1991), it is necessary that individuals be able to identify social and psychological benefits of reform efforts if they are to participate. Individuals must be able to identify the benefits of such movements and believe that movements will be able to achieve real benefits (Budd 2008). However, this does not necessarily mean that they must experience a crisis. Merely, that the benefit of reform must outweigh the potential costs. We believe that relying on a crisis might prevent or dissuade members from engaging in revitalization efforts. Therefore, we decided to explore whether successful reform efforts might indeed be initiated by something less than a crisis.

The Role of Democracy

If oligarchy is the problem that leads to the bureaucratization and stagnation of the labor movement (Voss and Sherman 2000a, 2000b; Michels 2009), then democracy is the solution. By definition, democratic movements to improve worker representation contrast with the efforts of oligarchic leaders to hold on to power. According to Labotz (1991 as cited in Moody 2007:176), rank-and-file activism is the foundation of the labor movement. According to Cook (1963), powerful unions tend to be active, and active unions are more democratic than inactive ones. According to Hochner (1980 in Barling, Fullager and Kelloway 1992), union democracy enhances union effectiveness by increasing power through increased union commitment and solidarity citing. Member involvement in decision making, a clear form of democracy, increases member participation as well as satisfaction (Bailing, Fullager and Kelloway 1992). Further, union power is built on power on the job (Moody 2007). As Parker and Gruelle describe, "Union power requires democracy" (1999 as cited in Moody 2007:176). Moody continues arguing Unions need active members to be strong and people will not remain involved for long if they don't have control over the union's program." (2007:176). Moody describes the undemocratic union in much the same way that Voss and Sherman described the bureaucratic, oligarchical union as a union moves "toward the passive professionally run AARP model of mass organization, you have a lobby with less and less influence, not a union with more power" (2007:176). According to Moody, the workplace is where unions have real power, but only if members have the power to influence how the union behaves: "If membership activity and involvement are discouraged, as is often the case, the union is weaker and the workers' position in the workplace becomes weaker still" (2007:176).

Dimick points out that British unions using generally more democratic processes than their American counterparts have been more successful in "traditional measures of union success" (2010:8). This suggests that the union democracy is a necessary but not a sufficient condition of union power. Nyden (1984) suggests that reform movements must focus, at least in part, on the positive, short-term incentive for members to have direct input in the reform movement (i.e., direct democracy). Without this democracy in a union, leaders and even staff members are likely to fall victim to the Michel's law of oligarchy.

The power of a union comes from the collective. As Dine (2008:228-229) writes,"[g]etting members more engaged in their unions is neither an inconvenience nor merely one possible strategy among others: it is a prerequisite to a revamped movement, a building block for whatever else labor hopes to do." The theory and the reality is that 100 people working together will always be more powerful than a single person on his own. However, to harness this power, those one hundred people must be involved and committed. To create this potential power, a union must be democratic—a necessary condition precedent to successful union reform. If union members do not believe that union leaders are working for them, then members will not support their unions (Barling, Fullager and Kelloway 1992); unions lose the fundamental source of their power—solidarity and concerted activity. Based on these, we hypothesize that union democracy will be a condition precedent to effective union reform. Further, the stronger the union democracy, the stronger the reform effort.

The Role of Militancy

The activation of the potential power of a democratic union is the next step in the model and that is the step of militancy. Militancy leads to the gains that drive revitalization, and it is at this step that our model comes back in line with Voss and Sherman. Moody (2007) points out that unions have grown during periods of militancy and shrunk during years of passive, servicing, cooperationist models of unionism. Moody (2007) points to the need for pressure and action that literally makes "business as usual" untenable for the organizations on a day-to-day basis. Bronfenbrenner and Hickey (2003) point out the importance of not only militant action, but an escalation of militant action as a part of a comprehensive organizing campaign.

We also see that where unions are not militant, members suffer (Moody 2007). Ackerman (n.d.) discusses how modern union leaders acquiesced to the current climate, accepting capitalism as the system and relying on government processes rather than strikes and other aggressive direct action as a way to maintain whatever control they could. Moody (2007) points to the example of the UAW ousting their own militant members and then suffering as a result when Chrysler was easily able to crush wildcat strikes in 1973, leading to a series of concessions from 1973-1985 by the UAW. Cooperationist movements such as the Kaiser Permanente and Saturn experiments have left those unions toothless in the wake of unilateral decisions by management and the closing of facilities. A lack of militancy or willingness to engage in militant actions has led to concessionary contracts and adoptions of the management-side business models. Danford, Richardson, Steward, Tailby, and Upchurch (2007) discuss how cooperationist unionism harms members, unions and the movement. A cooperationist approach cuts off the needs of the members as management needs are often more emphasized. Danford et al.'s (2007) case study of the aerospace industry shows similarity in the UK with increased privatization based on neo-liberal politics. In addition to more international supply chains, there was also large outsourcing of labor which weakened the union and sacrificed members' interests. Across a 10 year period starting in the mid 1990's, the number of employees in British Aerospace manufacturing was almost cut in half. Danford et al. (2007) also point out that union activists (stewards) were often targeted for job loss, and management may have had an easier time doing this according to the authors based on different seniority rules compared to the U.S.

The strategies of democracy and militancy complement and stimulate one another in what we suggest is a *virtuous cycle*. As mentioned earlier, a belief that a movement can be successful is a critical factor to driving involvement in that movement (Chong 1991). Successes of militant or aggressive actions will drive the perception of potential future successes, thus driving involvement. Also as stated earlier, the success of union tactics is predicated on collective power; as more members are involved, the tactics will be more powerful and more likely to be successful. The virtuous cycle will continue to build internal power.

This cycle will also build the external power. The practices of militancy and democracy have been shown to lead directly to organizing success (Brofenbrenner and Hickey 2003). Further, the shop floor successes that will result from these democratic and militant tactics will also lead to increased perceptions of union efficacy and can improve the general view of unions by clearly demonstrating that unions do have value for working people. These successes will lead to increased participation and increased desire to engage in aggressive tactics. Both the increased perceptions of efficacy and the desire to participate are critical factors in workers' decisions to vote for union representation (Wheeler 2002).

The Leaders' Response

Finally, we cannot ignore the union leaders' role in this process. Voss and Sherman express the issue of the leaders' roles by recognizing the pitfalls of oligarchy. There is little reason to expect that oligarchic leaders would accept a reform movement. Instead, entrenched union leaders will almost always work against democratic rank-and-file reform movements. American unions at their best are "nominally democratic" and are often "governed as one-party states" (Dimick 2010:4). According to Dimick,"Union leaders prefer oligarchy" (2010:5), and they will often take whatever steps necessary to protect the oligarchy, even if such steps harm or disenfranchise members or work against the goals of the broader labor movement. Instead of being a potential driver of revitalization, we suggest union oligarchs in all levels of formal leadership are often likely to be an inhibiting influence. In theoretical terms, union leaderships' responses moderate outcomes. Entrenched leader reactions seem to be overlooked in the Voss and Sherman model. Any prescriptive model for union reform must take into account the reactions seen from entrenched leaders. Reformers must be prepared to address and overcome these counterproductive reactions.

Testing the Model through Reform and Revitalization Case Studies and Efforts

From the above contextual cues and literature review, we developed a model of union reform that starts with some tipping point that leads to members engaging in reform efforts (see Figure 2). Once these efforts are undertaken, we hypothesize that the efforts will need to be aggressive (militant) in order to bring about the needed change and will also have to be democratic to bring a sustained support for such measures from the members of the union/labor movement. These democratic and militant efforts will have a synergistic effect upon each other, with militant efforts leading to successes which will lead to member willingness to engage, and this engagement in turn will allow the strength of numbers to strengthen the militant actions. Further, we hypothesized that rather than playing a supportive role, the entrenched leaders will be antithetical towards such a movement and will take measures to thwart these efforts. Again, to be successful, the reformers will have to respond to these leader responses in an effective way to minimize their damage to the reform effort. Our next step in exploring this model of reform was to review cases of union reform efforts to see if these themes did emerge.

The Tipping Points That Commence Union Reform Efforts

Voss and Sherman's model suggests that a political crisis in the union leads to a push for reform. However, within their model, they define this "crisis" very broadly. The crisis can include anything from concession bargaining, to "do nothing" leadership, to general chaos. More extreme crises such as union corruption, fraud, misappropriation of funds, and other unlawful acts would also be included. The key seems to be that there must be some tipping point that has pushed members to the point of demanding reform. Because of the broad definition of the crisis, we are most likely dealing with the same construct as the Voss and Sherman model, so we might simply diverge on semantics at this part of the model. However, it is important that we consider and discuss some details of what we have seen lead to the push for reform.

In our review, we found there to be common themes around the tipping points, as well as some variance. The reasons for the reform efforts included weak elected union leadership, concession bargaining, autocratic and/or corrupt leaders, cooperationist leaders, economic crises, and austerity measures. Our review of the tipping points in these various case studies, indicated that the tipping points can be categorized by the locus of the tipping point. We find that tipping points can be either internal (i.e., union operations especially in relation to members), or external, environmental factors that impact the union, or result from some combination of the two. We will further explore these three categories of tipping points in this section of our paper. Then we will summarize why we think the term tipping point best describes the impetus for reform.

Internal Tipping Points

We have found that in the case of the TDU, HERE under Sirabella, SMART, and Miners for Democracy, the tipping points were all internal. The Local 1010 of the UMWA began a push for reform within the mineworkers as reaction to leadership attempting to exert more control over members and ineffective handling of grievances by the leader (Feldman 1972). The Local 138 of the Teamsters led the push that would become the Teamsters for a Democratic Union as a result of substandard contracts and poor negotiating by the Teamster leaders, leadership attempts to suppress democracy, and the spread of corruption in the union leadership (Labotz 1990). SMART, the reform movement within SEIU, began as an effort to preserve the democracy that was present in many of the SEIU locals (Wulkan 2009). For the USW, the union leadership attempted to exert more control over members and this led to a reaction and push for reform (Nyden 1984). The Painters reform was a response to corrupt leadership, election stealing and rigging by union leaders, repression of dissent and business or even cooperationist unionism by union leaders at the expense of the members.

External Tipping Points

External, environmental factors also often push union members towards reform efforts. When political attacks or economic measures impact union members, this seems to spur members towards more active, militant and democratic activity within their union. The reform movement within RMT was a response against an ineffective union leadership response to privatization and its effects (Darlington, 2009). The Chicago Glassworkers revitalization was in direct response to a last minute plant shutdown and a refusal by the plant owners to pay the wages owed to the workers (Mead-Lucero, 2008).

Combination and Interaction off Internal and External Tipping Points

In many cases, we have found that the tipping point is either a combination of internal and external factors or a consequence of interaction between internal and external factors. As mentioned above, the CTU reform efforts came in reaction to austerity measures imposed from the outside. However, they also were a result of ineffective and weak reactions from union leaders to these external attacks. The USW Fight Back movement grew out of the increased control exerted by internal union members. However, the Fight Back movement also emerged in response to the first real downward economic pressure on the steel industry. Tipping points are complex and at times impossible to attribute to only internal or only external factors. The interaction between internal and external factors can serve as tipping points that spur union revitalization efforts.

Looking at internal and external tipping points as explanations for the initiation of revitalization efforts, it appears that "crisis" might be a fair characterization. However, at times it would seem that the push for reform might be more in line with a slow moving build-up of frustration—more of a process than an event. Gradual tipping points are often internal to the union (e.g., frustration with ineffective or apathetic leadership), but can be external (e.g., the RMT response to austerity measures), or a combination of the two (e.g., CORE's response to both austerity measures and the leadership's ineffective response to such measures). Table 1 provides a summary of the tipping points in the union revitalization efforts presented in this paper. This table includes a description of the tipping point as well as our analysis of the sources (internal, external or both) of the tipping point. (Note: All tables appear at the end of this paper, following the references.)

Tipping points may be attributable to discrete events, but others arise due to subtler processes. The impetus for change in the slow-building tipping points may be akin to low grade incivility or consistent patterns of member frustration building over time. So, the accumulated member dissatisfaction or patterns of leadership behavior can also be a tipping point. Clearly there must be some tipping point that transforms membership apathy into action. Apathy is a predictable outcome associated with oligarchical, unrepresentative, and/or stagnant union leadership. Tipping points may be predictors of the member activism and activity that is needed to initiate reform and revitalization efforts. We offer a new and broader conceptualization of crisis: *the tipping point of frustration*. We define this construct as discrete events or the accumulation of member dissatisfaction that stirs members to democratic action and militancy.

Democratic and Militant Strategies

The Voss and Sherman model of union reform efforts within the model of reform focuses on two sets of actors: (1) outside organizers who bring with them understanding and experience of other social movements and "alternative models of mobilization"; and, (2) and support from the international. While Voss and Sherman focus on the actors in the reform effort, they also mention the importance of the education or development of a culture of reform within the union. While we largely agree that these actors can indeed lead to union reform, we disagree with both the general position of outside organizers leading these movements and the need for international support. While we agree that outside organizers can play a vital role, and that international support would help to drive reform, we do not believe that either of these is necessary. Instead, we find that the success of union reform efforts is predicated on specific strategies and tactics rather than the actors who carry out the strategies and tactics.

Borrowing from strategic management terminology, we suggest that we should view the rank-and-file movement as an organization in and of itself. There is little doubt that for the movement to be successful it will need to act as a team with a single goal. We would again borrow from this strategic management terminology and define this goal as the vision of the reform movement. This vision would generally be consistent for all reform movements, and we could even argue that the reform and revitalization movement is one overarching organization. Whether at the level of union chapter, labor union, or labor movement, the mission is to revitalize to create better lives for working people, present and future.

Once reformers understand their mission and articulate their vision, the next step is to develop strategies of how to achieve these ends. We suggest that democracy and member militancy are the key strategies to reforming oligarchical, weak or stagnant unions. We argue that through strong (militant) measures that are focused on members' interests and driven by membership power, activism and mobilization, unions will become more successful in improving the lives of their members. These internal successes will drive member views of instrumentality and increase member satisfaction. We also present evidence to make the case that the activism and participation that is needed for these types of tactics to be powerful and successful can only exist in a democratic union where members have a say in the battles they wage.

In order to carry out the mission of union reform, there is a need for participation and militancy. In order to achieve the vision, militancy and democracy must be directed towards meeting the needs of working people through strong representation of members and through organizing efforts to increase worker representation. However, union apathy and lack of success has created a barrier to meeting these goals. In the next section of this paper, we present arguments for the strategies of democracy and militancy.

Democracy and Militancy

The Theme of Union Democracy

Many union reform movements place an explicit emphasis on democracy as part of the call for reform. We define *union democracy* as member-involvement and true member-driven decision-making. SMART, as mentioned earlier, is a direct result of looking to preserve democracy within the SEIU and its locals (Wulkan, 2009). Similarly, Teamsters for a Democratic Union is not only focused on democracy in its name, but also in its practices (Labotz, 1990). The Chicago Teachers Union is yet another example of a successful reform movement that focused on rank-and-file activism and democracy. The CORE reformers within CTU engaged in training rank-and-file members for leadership positions very early on in their movement (Bradbury, Brenner, Brown, Slaughter, and Winslow 2014). They also educated all of the members on issues and tactics and made a concerted effort to remain democratic. The same can be seen in Sirabella's and Wilhelm's successes at UNITE – HERE, with Wilhelm adopting Sirabella's faith in the members and the tactics of direct democracy. As Wilhelm is quoted as saying,

> I'll tell you things I think I learned from him (Sirabella) right at the beginning. One was that you had to be militant in order to get anywhere. Even more important was to have faith in the members. I have never to this day met another person, including myself, who has such endless faith in the workers. He absolutely believes that as long as you provide the workers with the facts and as long as you give them some leadership, they'll always do the right thing (Getman 2010:45).

Sirabella's position of faith in and reliance on members to make decisions epitomizes the type of democracy we are suggesting is needed for meaningful revitalization of any union.

Our review of the democratic practices of union revitalization efforts is depicted in Table 2. As can be seen in this table, most of the reform efforts contained at least some levels of democratic practices as drivers of the reform. In several of the reforms, such as SMART, TDU and the UMWA, democracy was not only a tactic or strategy of the reformers, it was also a goal of the movement. As mentioned above and below, labor researchers including Moody (2007) and Dine (2008) have suggested that democracy is needed to keep members activated, and active members are critical to union success. This is exemplified in settings where democratic processes allow unions, such as the Teamsters, to engage in successful, and often militant, direct action. The Teamsters have a history of top-down, oligarchical structure that began to change when TDU (Teamsters for a Democratic Union) began to gain influence within the Teamsters (Schiavone 2007). Eventually, the Teamsters' leadership under Ron Carey incorporated many of the TDU's principles into the Teamsters' organizational structure (more emphasis on day-to-day shop floor activity and union stewardship involvement, better education of members, etc.). These militant actions revitalized the Teamsters and led to the success of the UPS Strike in 1997 (Schiavone 2007). As a result, we should see reform movements that are able to implement democratic practices and will be more successful in terms of outcomes. In fact, the case of the Fight Back movement in the Steelworkers was less successful than it could have been because it failed to focus on truly democratic processes of allowing for immediate member involvement (Nyden 1984).

The Theme of Union Militancy

While our model diverges from the Voss and Sherman model in terms of how to get to strong union actions, we come back together in the need for militant action to revive a stagnant union and labor movement. We define *union militancy* as the willingness and ability of the union to engage in direct action and concerted activity that falls outside of the bureaucratic processes of collective bargaining and formalized, legalized, bureaucratized grievance processes under a CBA. Connor (2005) presents a good example of such militancy in the story of the ILWU calling for Harry Bridges whenever there is a dispute with a supervisor. The "call for

Harry Bridges" is actually a call to all union members to come to the support of union brothers and sisters by tangibly demonstrating solidarity and collective power.

Militancy also includes more traditional forms of union actions including informational picketing, nuisance grievances, recognition strikes, mass-protests, filing of ULPs, corporate campaigns, comprehensive organizing campaigns and strikes. These proven militant actions are successful in organizing the unorganized and gaining wins for the organized. Successful organizing and wins for workers are hallmarks of rank-and-file movements that revitalize stagnant or corrupt unions. We are not alone in our calls for greater militancy. The Voss and Sherman model is also predicated on these types of militant actions. Our research also provides evidence that militant actions are consistent elements of successful rank-and-file revitalization movements (see Table 3).

When a stagnant and weak Chicago Teacher's Union was ineffective at responding to austerity measures put into place by a Democratic Mayor, it was the CORE movement that responded through direct and militant actions that led to successes over many of the austerity measures (Bradbury et al. 2014). It was the CORE members' willingness to take over the roles that the CTU failed to perform that led to positive change in the union. CORE members represented members in grievances, held informational picketing, and helped members to file civil suits against the school district. These militant actions led to success on the substantive issues and led to a revitalization of the CTU under the new CORE leadership (Bradbury et al. 2014). Similarly, it was the militancy of the UE and their willingness to occupy their employer's premises that led to success for the Glassworkers in the same city (Mead-Lucero 2008).

Militancy is at the core of other rank-and-file reform movements as well. For instance, Ruscigno as one of the earliest leaders of the TDU movement demanded the right to file grievances on behalf of his members of the Local 138 (Labotz 1990). In the Mineworkers for Democracy rank-and-file movement, reformers at the Local 1010 actively filed grievances on behalf of members as a tactic for revitalization and recruitment (Feldman 1972). Sirabella's Unite Here is not only an example of the potential success via a democratic union, it also demonstrates the success of the willingness to use militant tactics (Getman 2010). Ackerman (n.d.) points out the need for militant action in order address the attacks on teachers and their unions. RMT (The National Union of Rail, Maritime and Transport Workers) in the UK is yet another example of the success of militancy and the need for unions to remain or return to a militant position (Darlington 2009). The CFDU (Campaign for a Fighting and Democratic Union) helped to revitalize RMT, and there were other loosely organized groups within RMT that also did this. RMT had other grassroots efforts throughout their history; perhaps RMT members were more accustomed to and comfortable with militant movements compared to other unions. RMT has a history of effective strikes, not only over pay, but also over management bullying and job cuts. Their militancy led to their breaking with the Labour Party, especially after Labour became more neo-liberal in their outlook and adopted more privatization of rail systems (Darlington 2009). Darlington argues that RMT's tactics have been beneficial for their members in terms of pay, benefits, and working conditions. At the time of the writing, RMT was one of the fastest-growing unions in the UK even as overall union membership declined. They adopt a "syndicalist" approach, antithetical to management and promoting re-distribution of wealth and power, rather than the more common "partnership" model that we see today (Darlington 2009). RMT's approach was a reaction to management's ever increasing emphasis on efficiency and productivity and the deals management would make with the union to reform work rules in exchange for pay increases.

An interesting point is made in the context of the Occupy Movement in 2011 and how weak unionism spills over into the way society views the labor movement. In the early stages, unions helped prevent the police from shutting down the protests. At the same time, many of the young radicals of the Occupy Movement tended to (apparently) see unions as ineffective tools for change in the workplace. Despite this exceptional case of support for broader social movements from some unions, union leaders tend to be more bureaucratic and conservative—making them reluctant to engage in important social movements (Ackerman n.d.). Reformers must oppose this weak, cooperationist type of unionism in order to revitalize the labor movement as a force for societal change. For rank-and-file reformers, this means they must take a page out the *Activist's Handbook* (Shaw 2001) and be willing to engage in the direct action and militancy that will lead to fear and loathing from the administrators and managers of organizations, as well as lead to the success, satisfaction, and increased perception of efficacy by members.

As described earlier, democracy and militancy create a virtuous cycle—working in concert to achieve revitalization. Militant actions are successful when they are carried out in number; a single militant leader cannot be effective without a large group of followers engaged with her lead. At the same time, a large group of individuals that is unwilling to engage in any type of pressure or militant tactic is unlikely to have any meaningful gains.

Theme of Union Leaders as Inhibiting Oligarchs

Voss and Sherman suggest that support from the International is a key to a successful rank-and-file reform movement. In this we disagree. While support from leadership would be an ideal situation, the reality is that such support is rare. More often, oligarchic leaders oppose reforms at every step. Despite these oligarchical reactions, the rank-and-file reform efforts can still be successful. However, what is needed is a ground swell of support from union members. Voss and Sherman accept that this piece of their model is antithetical to the usual social movement which tends to be bottom-up. We argue rank-and-file movements must not rely upon formal leadership support. The top-down, international support seems to suggest a model of corporate style unionism (Moody 2007). We suggest that a model of rank-and-file unionism is more likely to lead to meaningful internal and external reform. We believe a groundswell of support is critical. However, we also believe that militancy and democracy together create a virtuous cycle. Just as we see oligarchy and apathy create a vicious cycle. Thus, the democratic and militant tactics are likely to support each other and increase the rank-and-file support for the movement.

In the cases we have reviewed, entrenched leaders consistently take steps to maintain their power. The reactions from the entrenched leaders are presented in Table 4. These reactions are rarely focused on changing patterns and practices to meet the concerns and needs of the members pushing for reform. Instead, the reactions are antagonistic and defensive. These range from attempts to cast doubt on the veracity of the reformers, attempts to appease members by changing small practices or simply changing dialogues, to quashing the voice of reformers by policies banning campaigns, distribution of literature or tightening leadership control, to ostracizing, suing, threatening and even physically attacking reformers. For example, Teamster bosses stifled TDU reform candidates' efforts to be elected by using arcane clauses of the International Brotherhood of Teamsters (IBT) constitution (Goldberg 2005). Entrenched union leaders often tighten control over members, processes, meetings, and grievances. Oligarchs regularly use legal procedures, including suing their own members, to stifle reform movements. In some extreme cases, corrupt union bosses have engaged in violence against their own members and the reformers working to make their unions better.

While union reformers have the goals of improving their unions and opening the union to the voices of *all* members and interests, they must be aware they will rarely convince entrenched leaders of the need for change. While the CTU reformers did recruit some of the entrenched leaders into the operations of the CTU, this did not occur until after the reformers had taken office (Bradbury et al. 2014). The pattern for entrenched leaders is to take whatever actions are necessary to hold onto their personal power, rather than strive to improve their union.

Reform movements must be able to respond to and counteract these actions in order to survive. At this stage, rank-and-file reform efforts often become battles between the reformers and entrenched oligarchs. The reformers struggle to garner enough support to have a meaningful impact on their union, while the entrenched leadership seeks to thwart reform and maintain the status quo. Reformers should select effective strategies, processes, and tactics to overcome these reactions (Nyden 1984). Some strategies include focusing on the reformers' strengths, defining and exploiting areas of member disenchantment, and building reform leaders' ability to influence fellow members. All of this will have to be a conscious action on the part of reform leaders and the members who follow them. As mentioned earlier, successful revitalization efforts are grassroots movements. They are led by rank-and-file members and provide opportunities for the rank-and-file to be directly involved in the short term and long term objectives (Nyden, 1984). Reform leaders must influence and persuade union members and combat the messages from the entrenched leaders. At this stage, the reform movement often is a matter of persuasion and mobilization. In essence, this entails mobilizing members against entrenched leaders—the movement must clearly articulate the injustice of the leadership action (or inaction) and relate this to negative worker outcomes (Martin and Sinclair 2001). As Dimick (2010) points out, mobilization of the rank-and-file is necessary to remove the old guard. Another virtuous cycle emerges as persuasion leads to mobilization, leads to democracy, leads to militancy outcomes, leads to persuasion, leads to mobilization, *ad infinitum*. To foster this virtuous cycle, reform leaders must take steps to assure they are effective in promoting member interests and are accessible, reliable, and responsive to member demands (Schutt 1982). This is illustrated by the Chicago Teachers reform when they addressed member concerns from the first day including helping members to file civil discrimination suits (Bradbury et al. 2014). Further, it is common for reform leaders to emerge from stewards who are already taking steps to address member interests. Other reformers can be activated through social ties and personal contacts (Kuruvilla and Fiorito 1994; Fiorito et al. 2011). Finally, the movement must convince members of the collectivist nature of their issues (Buttigieg, Deery, and Iverson 2008) via transformational leadership (Kelly, 1998). Examples of transformational union leaders include Sadlowski with Fight Back, Yablonksi with the Mine Workers and, of course, Sirabella with Unite HERE.

The Tactics of Democracy and Militancy

As action researchers, we hope that our work will help to guide practices. Therefore, we think it makes sense to go beyond mere strategies of reform movements to discuss tactics that are used to carry out these strategies. By delving into the specific tactics, our hope is that union leaders and reformers will be able to adopt tactics that have proven to be successful in other reform efforts or adapt these tactics to fit their own efforts.

Tactics of Democracy

Democracy in these reform efforts is more than simply having a mechanism for members to vote for leaders. Democracy is not just about structures and processes, it is about "developing a culture of activism, involvement, transparency in union governance, and openness in debate" (Moody 2007, p. 178). Even the most oligarchical unions have these types of mechanisms in place. Instead, the tactics must dig deeper and focus on a broader concept of democracy. The tactics for union democracy can be divided into three categories—union elections, members' education, and shop floor rank-and-file democracy and activity.

In perhaps the earliest study of union democracy, Lipset and colleagues (1956) concluded that the ITU was a strong democratic union that had avoided the Iron Law of Oligarchy because the union had a history of secret societies, the union had two strong parties with clearly contested elections, the membership was well educated, and the different parties established connections in the shop but also through social gatherings. The Association Union Democracy (AUD) supports all of these factors as important to union democracy but also adopts further practices including frequent, fair, and contested elections. All candidates must have access to membership lists; to assure union democracy, campaigns should be fair and open, and members must be educated to make truly informed electoral decisions (Democracy 2011). The AUD also suggests that in order to be democratic, members must have true rights in all phases of the labor relations process including full information about collective bargaining proposals and agreements, the full right to vote on the CBA, and rights to manage and be involved in their own grievance process (Democracy, 2011).

From the standpoint of the union reform movement, the ability to focus on and effectively resolve workers' day-to-day interests and direct concerns seems to be a key. Dimick (2010) suggests that informal shop floor associations are a key pathway to this linkage. For Moody, the link is also a workplace association, but it can be formal or informal. In many cases, the association will be a "stewards' organization that is directly responsible to the members in the workplace" (Moody 2007:178). A reform party was present in most of the instances of reform movements we reviewed including Lipset's ITU, the Painters, the CTU, the Mineworkers, the USW, the Teamsters and the SEIU. The possible exception to this was Sirabella's UNITE-HERE; even in that case, the opposition party was clearly recognized if not formalized. Many of these parties still exist today. Some of the reform parties we identified are Miners for Democracy, SMART, Fight Back, and the Teamsters for a Democratic Union (TDU).

Moody (2007) sets out a number of conditions that must exist (or be avoided) in order to accomplish union democracy. The grievance process must not be overly bureaucratized or legalized to the point of being inaccessible and ineffective. The path towards union democracy is one of democratic social justice unionism rather than traditional business unionism or the Stern and Lerner (year) model of bureaucratic unionism. Moody's three principles that must exist to assure union democracy are:

1. Power must be in the workplace at the workplace level.
2. True internal union democracy. Power comes from member participation and in order to make members want to be involved, true democracy must exist; and, the members must be able to be directly involved with issues and the union power must be deep in the workplace.
3. Unions need to reach out to other justice based organizations to form alliances and build power (Moody 2007).

We see these tactics utilized throughout successful reform movements. The key ingredient for carrying out the strategy of democracy in the workplace is the presence of direct democratic practices. Members should participate directly in decisions via union meetings. If the practices or processes are based on representative democracy, the representatives should be elected directly by and be responsible to the members. RMT exemplifies a bold style of democracy. Every three years the Executive Council of RMT must step down (Darlington 2007). Their annual meetings have a similar structure in that delegates cannot attend these for more than three consecutive years, "encouraging a high degree of devolved activist engagement" (Darlington 2007:19). RMT also frequently engages members with consistent communication such as regular newsletters, emails, and even text messages. RMT's activism has boosted members' self- confidence (an upward spiral rather than downward one)—this also led to increased membership after declines. Union reps in RMT are clearly more empowered and carry greater responsibilities.

Successful reform movements focused on openness and participation from members and were committed to remaining democratic while the entrenched leaders remained oligarchic. In the TDU, the UMWA, Unite-HERE and the CTU there was also a clear focus on the members' day-to-day issues and interests. These rank-and-file movements took on individual claims for grievances and even lawsuits. Rusigno filed grievances for his members even when leadership refused to provide grievance forms despite receiving physical threats (Labotz 1990). The CORE leadership took steps to help members in the filing of their civil rights suit well before they had any formal positional authority (Bradbury et al. 2014). Successful movements kept the day-to-day needs of rank-and-file members at the forefront of their action agenda. Revitalization leaders trusted their union brothers and sisters and relied on mere members to lead the way. The CORE leaders wisely trained rank-and-filers to be ready to step into leadership roles. All of the movements gave members the chance to have greater say in the actual decision making process in their unions.

Tactics of Militancy

Much like democracy, we must understand what we mean by militancy. Militancy is not just theory, it is action-oriented. Traditional militant tactics like strikes, walk-outs, slowdowns, and picketing are fairly clear examples within the labor movement. In this case, we are suggesting that there must be ability and commitment to engage in these actions. A truly reformed union will engage in militant tactics when the bureaucratic systems of collective bargaining and the grievance process do not lead to gains for their members.

Ackerman (n.d.) argues that militant actions would have to come through informal organizations and only through information organizations. He suggests that "a militant current within the existing unions could support the creation of independent worker organizations possessing no assets and no property. These organizations would be able to violate Taft-Hartley and other laws: to strike and organize using tactics that defy the authorities and target the shutdown of production without fear of losing years of accumulated strike funds in lawsuits or court fines."

We cannot ignore the fact that many militant actions of the past were not protected under the NLRA or state labor relations laws (just as they are not protected today). In fact, the penalties were harsher than what we see today. Thus, we know that militant actions often do come through formal union structures and process, but they also have potential costs. Some militant tactics are protected under the NLRA and state labor laws; these fall outside of and go beyond the grievance and collective bargaining process. Even within these formal processes, unions can undoubtedly be militant. Economic strikes are in fact protected under the NLRA. Despite this protection, the flow of economic strikes we saw beginning after World War II has substantially reduced. While there were over 300 major strikes per year in much of the 50's, 60's and 70's, this number has dwindled to just a handful per year during the past decade (Rosenfeld 2014). Does this lack of strike action indicate that union workers are getting better contracts and have no need to strike? Anything but. In fact, during this time the standard form of bargaining has become concessionary bargaining (Moody 2007). Acceptance of the employer argument that we are all in the same boat and unions must negotiate with the employers' bottom line in mind has led to a period of stagnating wages and booming profits (Mishel, Bivens, Gould, and Shierholz 2012).

We also know that grievances are not only a protected form of militant action, but they are also a fully accepted part of the labor relations process and an important clause negotiated in most CBAs. However, in many of the unions we reviewed, union leaders refused to file grievances on behalf of their members. Grievances can follow rigid formal steps; union leaders can limit grievances by pursuing only clear and substantial CBA violations. What many unions fail to realize today is that grievances can also be employed as a militant tactic. Grievances can be filed to fire a warning shot, shine a light on problems, build a record, or to forge employee unity (Mauer 2001). Mauer (2001) also points out that the method of filing a grievance can be used to increase direct pressure. For instance, a series of grievances can be filed by each individual affected by the same employer decision, or a number of members can sign off on a single grievance. Conner (2005) and Mauer (2001), in other circumstances, argue that other forms of direct action may be more effective than filing grievances. Mauer (2001) presents the example of all workers refusing to buy food from the company cafeteria when the company decided to outsource the cafeteria jobs. As previously mentioned, Conner (2005) presents the example of the ILWU "call to Harry Bridges" as an effective and militant direct action.

Militant actions can also occur outside of these two processes. The NLRA, and many state labor laws, protects "concerted activity for mutual aid or protection" (NLRA 1935). This language is not limited to collective bargaining and filing grievances. The direct action mentioned above in lieu of the filing of a grievance is one such example. It is also clear that picketing such as that by low-paid workers against McDonalds and Wal-Mart workers is also protected activity under the NLRA (Douglas 2014). The NLRA provides procedural rights to engage in the formal processes of collective bargaining and grievance filing, but it also provides union members the right to use their economic and collective power for their mutual aid and protection. There are many protected forms of militant actions under the NLRA and state labor laws.

However, a revived labor movement cannot only make decisions based on legality. There has been no successful social movement in the United States that has not seen members of the movement at times step outside of the bounds of the law. In fact, the laws (which are written by those in power) are often set up to hinder the success of social movements. One of Martin Luther King's most famous letters was indeed written from a jail cell. The Freedom Riders broke unjust laws in Alabama, Mississippi, and other southern states. The Pennsylvania teachers unions of the 1960s were barred from striking yet, this was the heaviest period of teachers' strikes in the history of the state (Shaffer 2011). Today we continue to see examples of innovative reformers taking actions that fall outside of the law. Strikes by public sector workers in Michigan continue to be banned, yet the faculty at Eastern Michigan continues to be willing to engage in strike actions if they do not receive a fair contract (Capriccioso 2006; Washburne 2006). Those involved in the Occupy movement were willing to face arrest to bring attention to the growing gap between the 1% and the rest of us (Wall Street Protestors: We're in it for the long haul 2011). The Chicago Glass workers were willing to violate trespassing laws by occupying their employer's premises (Mead-Lucero 2008). Protestors standing for better wages for fast food, retail and other low wage workers have been arrested and jailed for their actions as have those protesting police brutality (Gardner 2014; Greenhouse 2014). Militant action is not always clean, it is not always comfortable, but it is often effective and necessary to achieve real gains. It is difficult to find a case where those in power ceded their privilege without a struggle. Instead, the success of any social movement depends on the movement creating fear and loathing from those in power (Shaw 2001).

Within the labor movement, we can no longer afford to be swayed by leaders who warn of the dangers of taking direct action, engaging in strikes, or harming the working relationship with management by filing grievances. We cannot watch the gains of CBAs-past be forever lost and passively accept the decline of the broader labor movement. We must not forget the many societal benefits brought about by the collective power of unions and workers. In order to revitalize the labor movement, we must revitalize individual unions.

In the following sections of this paper, we will explore the outcomes associated with successful revitalization movements and use the DeMRev model to explain these outcomes.

Outcomes of Successful Reform and Revitalization Movements

When reform movements focus on the strategies of democracy and militancy, they mobilize their membership to engage. To become democratic militant unions, they create the necessary conditions to achieve union revitalization. Whether or not a union reform effort is successful often depends on the amplitude and duration of the oligarchic response and the success or failure of the effort to address and overcome the reflexive resistance of entrenched union leaders. When reform movements successfully revitalize their unions, members expect to achieve real gains. This brings us to the question of what this type of unionism will achieve. Goldberg (2007-2008) argues persuasively that democracy (and perhaps militancy) might have value in and of themselves. However, the focus of our investigation is the pragmatic benefits accruing to democratic, militant, revitalized unions. Such a union will not sacrifice member interests in order to remain in power because *members* of such a union hold the power. As mentioned earlier, the democratic union creates a virtuous cycle in which democracy leads to greater involvement, greater efficacy, and more militancy. The union that is effective at meeting member interests will lead to more members being active, committed, and satisfied (Bariling, Fullager, and Kelloway 1992). Participative leadership, support for members' direct participation, and freedom to voice opinion are all elements likely to exist in a union that has undergone successful reform and revitalization—all predictors of union satisfaction (Bariling, Fullager, and Kelloway 1992). Shop floor militancy will make it more difficult for management to force the union into concessions (Moody 2007). There is also clear evidence of unions growing during years that they are militant and shrinking in years that they are not (Moody 2007). Maranto and Fiorito (1987) find that more democratic unions have better organizing success, concurring with the Bronfenbrenner and Hickey's (2003) work on comprehensive campaigns.

We present revitalization outcomes in the case studies we have reviewed as indicated in Table 5. Positive outcomes range from comprehensive victories to modest changes. CORE's revitalization of the CTU and the MFD ouster of criminal leaders of the UMWA are examples of comprehensive victories. Other outcomes are incomplete or short-lived. Victories may occur at only one level or affect only one local. The reformers of SMART and Local 2, HERE did not succeed in revitalizing their entire union, but garnered dramatic and long-term changes at the local level (Wulkan 2009; Getman 2010). Both the Painters and the TDU succeeded in a series of elections to remove entrenched leaders. Other "unsuccessful" efforts, such as Fight Back, pressured resilient oligarchs into being more responsive to member-driven issues such as fair wages. Other partial successes such as at the NYPPU resulted in temporary revitalization (Hall 1972). The NYPPU also serves as a cautionary tale as to the fragility of revitalization efforts. A revitalized union will only remain effective as long as it remains member-driven, democratic, and militant. Today's reformers can become tomorrow's oligarchs if they allow their unions to fall out of the virtuous cycle of democracy and militancy. On the other hand, the RMT maintains revitalization and avoids oligarchy through its consistent democratic processes and commitment to militancy.

The CORE reform of the CTU led to a successful contract as well as a social movement that pushed back many of the most damaging austerity efforts in the Chicago School District (Bradbury et al. 2014). The TDU reform, in addition to other successes, played a large part in the successful UPS strike of 1997 (Moody 2007; Schiavone 2007). UNITE-HERE under Sirabella and Wilhelm is one of the few examples of true organizing success in modern era (Getman 2010). The RMT was consistently successful at organizing new members and achieving gains from direct action in a difficult political environment (Darlington 2009). The revitalized Chicago Glassworkers used direct action to save their jobs and facilitate a worker buy-out of their manufacturing facility (Meade-Lucero 2008). MFD succeeded in removing corrupt leaders revitalizing the UMWA; this revitalization refocused the UMWA on its mission to protect its workers (Feldman 1972).

The Contribution of the DeMRev Model

Revitalization of stagnant and/or oligarchical unions is indeed possible. In 2000, Voss and Sherman presented evidence of their model of union reform from above. Fourteen years later, we propose a model of bottom-up reform that we believe is more sustainable and holds greater potential for a lasting reform movement. Depicted in Figure 2, the DeMRev model focuses on reformer actions required to gain the grassroots members' support needed to truly become a movement (both inside their own union and more broadly as part of the labor movement). The first step of this model concerns tipping points. Tipping points provide the impetus for change within a union. Reformers who capitalize on tipping points can work to create a broad-based grassroots movement which replaces the apathy and complacency among members so typical of oligarchic unions. Tipping points cause members to perceive more pain in doing nothing than in making efforts to take on issues. In some cases, this tipping point might result from crisis. The push for austerity in the Chicago School District presented such a crisis that led to the CORE efforts. In other cases, tipping points might result from the gradual loss of benefits due to weak CBAs or the gradual increasing of control or declining representation by union leaders such as we see in the HERE, MFD and Fight Back movements for reform.

Once a tipping point occurs and a revitalization effort is achieved, we suggest that the movement must focus on strategies of democracy and militancy and implement concrete action-oriented tactics to make a movement democratic and militant. Democratic strategies and tactics are present in many of the reform movements we have reviewed including SMART, HERE, CORE and the MFD. Not only must the reform movements be democratic, but they must also be militant. Again we see militant activity throughout successful reform movements. The TDU focused on filing grievances and pushed the Teamsters towards the willingness to engage in the UPS Strike. CORE led to the CTU teachers' strike.

Figure 2
DeMRev Model for Union Revitalization

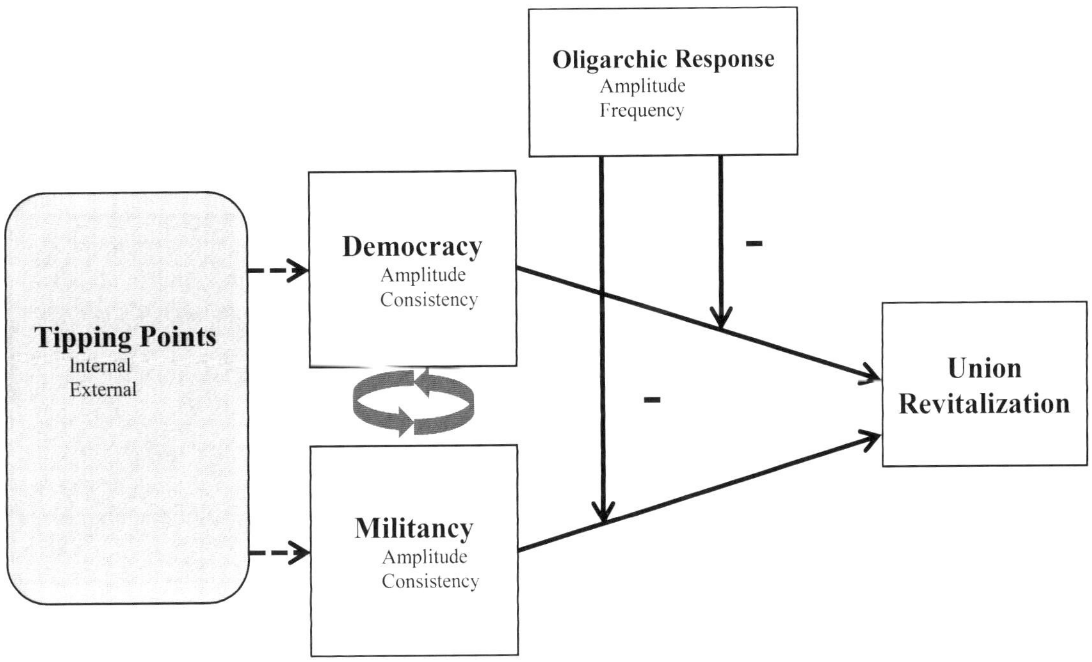

The leaders of Fight Back, MFD, and Sirabella's HERE engaged in militant activities which achieved real gains for their members. We suggest that not only are these strategies and tactics important, but the amplitude andfrequency of these tactics are important. For successful, revitalized unions such as the RMT, democracy and militancy are not brief experiments in reform, but consistent long-term practices strongly embedded in the culture of the union. In the case of RMT, they developed policies and bylaws which ensure that members drive the union and that leaders cannot become entrenched. In other successful reform movements, such as CORE and Sirabella's HERE, full involvement of members and true democracy existed throughout the movement, in every tactic and at every time. In addition, CORE and HERE were willing to engage in militant activity whenever needed. The DeMRev model explains the success of RMT, CORE, and HERE because their militant and democratic activities reinforce one another; together, democracy and militancy create a virtuous cycle which can sustain and propel a revitalization movement. When members are involved in decision-making, such as the decision of Chicago glassworkers to engage in a sit-in, they are more likely to be willing to engage in militant activity; when militant activity leads to successes, workers will be more likely to engage in union activities to continue to fight for gains. The Fight Back movement perhaps presents a counterexample. While leaders of this movement were indeed willing to engage in militant actions at the shop floor and beyond, the movement lacked a focus on direct democracy. Ultimately, reform efforts seem unlikely to succeed without a sustained commitment to both democracy and militancy.

These democratic and militant strategies and tactics will help the movement to gain support, but they will also of course draw the attention of the entrenched oligarchs. We suggest that these oligarchs and their reactions to the reform movements will play the role of inhibitor or moderator of the reform movement. Much like the amplitude and frequency of the reformers efforts, the oligarchs' amplitude and frequency will determine the level of moderation of the reform movement. The Painters' oligarchs seem to have been able to wait out the reform movement. The oligarchic responses did not stop even when the oligarchs were temporarily displaced. Eventually the oligarchs' persistence led to a restoration of their power. The reform movement's ability to overcome these oligarch reactions will be critical to the success of the reform movement. Where both sides are persistent and the amplitude and frequency of their activities remains high, we often see the battles drag on between reformers and the oligarchs. We continue to see a struggle between SMART and the entrenched leaders in the SEIU, and the TDU and the entrenched leaders in the Teamsters. Even in the CTU it is perhaps too early to determine the complete moderating effect of the responses from the oligarchy and whether we will see renewed efforts from these former leaders.

Conclusion

The current labor movement is clearly in desperate need of reform and revitalization. With union density around 11% overall and a mere 7% in the private sector, this is indeed a low point for organized labor in the USA (Rosenfeld 2014). In addition, we see (not coincidentally), that the American worker is struggling, with work intensification, stagnating wages, and growing numbers of low wage and poverty-wage workers (Rosenfeld 2014). As the labor movement has weakened, we have seen concession bargaining become the common practice while direct action has become a thing of the past (Moody 2007; Dine 2008; Yates 2009; Rosenfeld 2014). Union and non-union workers, as a result, have both suffered over the last several decades (Mishel, Bivens, Gould, and Shierholz 2012). A labor movement that will address the issues facing the working people of the United States is vital if we are to achieve social justice. We must provide the tools to revitalize the labor movement one union at a time.

Union revitalization *is* possible, and there are many successes that help us to understand the path to reform. By reviewing some of these successful rank-and-file reform movements, we have presented one potential model for union revitalization. While there are many factors external to the labor movement including globalization, expansion of union busting, weakening of labor laws, and economic crises caused by destructive capitalism (Carbo, Langella, Dao, and Haase 2014), unions cannot blame only external factors on their grave situation. Internal union factors such as business unionism, cooperationist unionism, entrenched oligarchic leadership, and the resultant member apathy present serious challenges to the labor movement. Rather than blaming all union setbacks on hostile environments, we in the labor movement must take responsibility to clean our own houses. Only with revitalized unions can the broader labor movement succeed.

In this article, we present a strategic model of member-driven reform that can be initiated by rank-and-file reformers and progressive leaders within unions and the broader labor movement. At the end of the day, the success of such reform movements will depend on broad member support, so the reform must come at a point when members are ready and unafraid. Members must be frustrated enough with current union leadership or current union practices that they are willing to support reform. From there, the reform movement must adopt the strategies of union democracy and union militancy to both gain support and to revitalize the union. There are proven tactics to encourage member involvement and shared decision-making. Case studies provide evidence that focusing on the day-to-day issues of members and engaging in direct action, grievances, and protests can make a difference. Finally, the rank-and-file movement must be prepared to respond to varied reactions from the all-too-present oligarchs who run so many unions. Reformers must respond by maintaining a focus on the members and rewarding member efforts with concrete gains. The road to revitalization is difficult and fraught with obstacles. It's not an easy path to follow. Revitalization does not come for free. Meaningful positive change never does.

References

Ackerman, S. (n.d.). *The Strike and its enemies.* Retrieved January 9, 2015, from Jacobin: https://www.jacobinmag.com/2011012-the-strike-and-its-enemies/

Barling, J., Fullager, C., & Kelloway, E. K. (1992). *The Union and Its Members: A Psychological Approach.* New York AND Oxford: Oxford University Press.

Bradbury, A., Brenner, M., Brown, J., Slaughter, J., & Winslow, S. (2014). *How to Jump Start Your Union: Lessons from the Chicago Teachers.* Detroit, MI and Brooklyn, NY: Labor Notes.

Brofenbrenner, K. (2005). What is Labor's true Purpose?: The Implications of SEIU's Unite to Win Proposals for Organiznig. *New Labor Forum, 14*(2), 19-26.

Brofenbrenner, K., & Hickey, R. (2003, June). Winning is Possible: Succesful Union Organizing inthe United States - Clear Lessons, Too Few Examples. *Multinational Monitor*, 9-14.

Buttigieg, D. M., Deery, S. J., & Iverson, R. D. (2008). Union Mobilization: A Consideration of the Factors Affecting the Willingness of Union Members to Take Industrial Action. *Britsh Journal of Industrial Relations, 46*(2), 248-267.

Capriccioso, R. (2006, September 5). *Strike at Eastern Michigan.* Retrieved January 8, 2015, from Inside Higher Ed: www.insidehighered.com/new/2006/09/05/strike

Chong, D. (1991). *Collective Action and the Civil Rights Movement.* Chicago: The University of Chicago Press.

Connor, M. A. (2005). Harry Bridges Prowls the Stacks at Powell's. In J. e. Slaughter, *A Troublemaker's Handbook: How to Fight Back Where you Work and Win* (p. 13). Detroit, MI: Labor Notes.

Cook, A. H. (1963). *Union Democracy: Practice and Ideal.* Ithaca, NY: Cornell University Press.

Danford, A., Richardson, M., Stewart , P., Tailby, S., & Upchurch, M. (2007). Capital Mobility, Job Loss, and Union Strategy: The Case of the UK Aerospace Industry. *Labor Studies Journal, 32*, 293-318.

Darlington, R. (2009). Leadership and Union Militancy: The Care of the RMT. *Capital & Class, 33*(99), 3-32.

Democracy, T. A. (2011, February 8). *AUD's Union Democracy Benchmarks.* Retrieved January 14, 2015, from AssociationforUnion Democracy: http://aud2.uniondemocracy.org/auds-union-democracy-benchmarks-0

Dimick, M. (2010). Revitalizing Union Democracy: Labor, Law, Bureaucracy, and Workplace Association. *Denver University Law Review, 88*(1), 1-56.

Dine, P. M. (2008). *State of the Unions: How Labor Can Strengthen the Middle Class, Improve our Economy, and Regain Political Influence.* New York: McGraw Hill.

Douglas, D. (2014, January 16). NLRB charges Wal-Mart with violating worker rights during employee protests. *The Washington Post*, p. A16.

Feldman, C. R. (1972). Miners for Democracy. In B. Hall, *Autocracy and Insurgency in Organized Labor* (pp. 11-21). New Brunswick, NJ: Transaction Books.

Fiorito, J. T. (2004). Union Renewal and the Organizing Model in the United Kingdom. *Labor Studies Journal, 29*(2), 21-53.

Fiorito, J., Tope, D., Steinberg, P. E., Padavic, I., & Murphy, C. E. (2011, December). Lay Activism and Activism Intentions in a Faculty Union: An Exploratory Study. *Labor Studies Journal, 36*(4), 483-507.

Gardner, B. (2014, December 11). Arrests as Protestors Stage Mass "Die-in" at Westfield Shopping Centre. *The Telegraph.*

Getman, J. G. (2010). *Restoring the Power of Unions: It Takes a Movement.* New Haven and London: Yale University Press.

Goldberg, M. J. (2007-2008). In the Cause of Union Democracy. *Suffolk University Law Review, 41*, 759-771.

Goldberg, M.J. (2005). Teamster Reformers: Their Union, Their Jobs, Their Movement. *Journal Transportation of Law Logistics and Policy, 72*:13.

Greenhouse, S. (2014, September 5). Hundreds of Fast-food Workers Striking for Higher Wages Arrested. *NY Times*, p. B3.

Kelly, J. E. (1998). *Rethinking Industrial Relations: Moibilization, Collectivism and Long Waves.* London: Routledge.

Kuruvilla, S., & Fiorito, J. (1994). Who Will Help? Willingness to Work for the Union. *Industrial Relations*, 548-575.

Labotz, D. (1990). *Rank and File Rebellion: Teamsters for a Democratic Union.* London and New York: Verso.

Lipset, S. M., Trow, M. A., & Coleman, J. S. (1956). *Union Democracy: The Internal Politics of the International Typographic Union.* Glencoe, IL: The Free Press.

Maranto, C. L., & Fiorito, J. (1987, January). The Effect of Union Characteristics on the Outcome of NLRB Certification Elections. *Industrial and Labor Relations Review, 40*(2), 225-239.

Martin, J., & Sinclair, R. R. (2001). A multiple motive perspective on strike propensities. *Journal of Organizational Behavior, 22*(4), 387-407.

Mauer, M. (2001). *The Union Member's Complete Guide: Everything you want - and need - to know about working union.* Annapolis, MD: Union Communication Services, Inc. .

Mead-Lucero, J. (2008, December 22). Chicago Sitdown Stike Produces Win for Workers, not Banks. *Labornotes.*

Michels, R. (2009). *Political Parties: A sociological study of the oligarchic tendencies of modern democracies.* New Brunsick, NJ: Transaction Publishers.

Milkman, R., & Voss, K. (2004). *Rebuilding Labor: Organizing and Organizers in the New Union Movement.* Ithaca, NY and London: Cornell University Press.

Mishel, L., Bivens, J., Gould, E., & Shierholz, H. (2012). *The State of Working America* (12th ed.). Ithaca, NY: ILR Press.

Moody, K. (2007). *US Labor in Trouble and Transition: The Failure of Reform from Above, the Promise of Revival from Below.* London and New York: Verso.

Nyden, P. W. (1984). *Stellworkers Rank-and-File: The Political Economy of a Union Reform Movement.* South Hadley, MA: Bergen and Garvey.

Rosenfeld, J. (2014). *What Unions No Longer Do.* Cambridge, MA: Harvard university Press.

Schiavone, M. (2007). Rank-and-file militancy and power: Revisting the Teamster struggle with the United Parcel Service ten years later. *The Journal of Labor and Society, 10*, 175-191.

Schutt, R. K. (1982, April). Models of Militancy: Support for Strikes adn Work Actions Among Public Employees. *Industrial and Labor Relations Review, 35*(3), 406-422.

Shaffer, R. (2011, Augudy). Public Employee Unionism: A Negelect Social Movement of the 1960s. *The History Teacher, 44*(4), 489-508.

Shaw, R. (2001). *The Activist's Handbook: A Primer* (2nd ed.). Berkeley, CA and Los Angeles: University of California Press, Ltd.

The Merriam-Webster Dictionary. (1974). New York, NY: Pocket Books.

The National Labor Relations Act. (1935). *Section 7.*

Voss, K., & Sherman, R. (2000, September). Breaking the Iron Law of Oligarchy: Union Revitalization in the American Labor Movement. *American Journal of Sociology, 106*, 303-349.

Wall Street Protestors: We're in it for the long haul. (2011, October 3). *The New Zealand Herald.*

Washburne, T. W. (2006, September 11). *Teachers' Strikes, Court Orders and Michigan law.* Retrieved January 8, 2015, from Mackinac Center for Public Policy: www.mackinac.org/7922

Wheeler, H. N. (2002). *The Future of the American Labor Movement.* Cambridge, UK and New Yourk: Cambridge University.

Wulkan, F. (2009, May-June). Quest for Democracy Persists Inside SEIU. *Union Democracy Review*(179). Retrieved January 7, 2015, from http://www.uniondemocracy.org/UDR/193-Quest_for_democracy_persists_inside_SEIU.htm

Yates, M. D. (2009). *Why Unions Matter* (2nd ed.). New York: Monthly review Press.

Table 1
Tipping Point Initiating Revitalization Efforts

Union Revitalization Context	Tipping Point	Locus of Tipping Point
Caucus of Rank & File Educators (CORE) Chicago Teachers Union (CTU) Chicago, Illinois, USA 2010-present	Privatization of public education Political attacks on public schools Weak CTU response to challenge	Internal and external
Teamsters for a Democratic Union (TDU) Teamsters International USA and Canada 1976-present	Leadership consolidation of power Top-down decision making Corruption within union Contract negotiation failure	Internal
National Union of Rail, Maritime and Transport Workers (RMT) United Kingdom Late 1990s to early 2000s	Privatization of London Underground	External
Fight Back United Steel Workers USA 1960s	Economic pressure on industry Management refusal to grant wage increases Overreach of leadership with regard to control of union functions	Internal and external
Local 2, Hotel Employees and Restaurant Employees (HERE) New Haven, Connecticut, USA 1960s-1970's	Unions abandonment of class struggle Business unionism Top-down decision making Association with Sirabella	Internal
SEIU Member Activists for Reform Today (SMART) Service Employees International Union (SEIU) USA and Canada 2008-present	Leadership consolidation of power Top-down decision making Business unionism "Sweetheart" deals & secret negotiations Abandonment of long-term goals for short-term gains, e.g. building worker power	Internal
Chicago Glassworkers 2008 Chicago, Illinois, USA	Plant shutdown with three day notice from owners	External
UNITE HERE Hotel Employees and Restaurant Employees (HERE) USA 1960s-1970's	Leadership return to business unionism Leadership return to top-down decision making	Internal
Miners for Democracy (MFD) United Mine Workers of America (UMWA) USA 1969-1970's	Corrupt static leadership Business Unionism Leadership ignoring worker safety issues Continual negotiation failures Financial corruption Murder of union reform leader	Internal
Rank-and-file movement, Active Painters Club New York Painters and Paperhangers Union (NYPPU) New York, USA 1930s-1960s	Association with organized crime Corrupt leadership Election Stealing Business unionism Repression of dissent Special treatment of part of the membership Investigation by law enforcement	Internal and external

Table 2
Democracy as a Strategy for Revitalization

Revitalization Effort	Democracy
CORE	Made union events social events Created candidate slates to run in opposition to leadership Contested CTU elections Worked to solve meeting attendance issues Focused on transparency of processes Created candidate slates to run in opposition to leadership
TDU	Contested teamster elections Built networks connecting members to each other Forced Teamsters to improve democratic processes Forced Teamsters to respond to member concerns
RMT	Continued with democratic and transparent union elections Adopted rules and bylaws that ensured turnover in top leadership
Fight Back	Built networks connecting members to each other Emphasized personal contact on the job Created candidate slates to run in opposition to leadership Contested USW elections
Local 2, HERE	Changed committee membership from appointments to elections Explicit and genuine faith in membership concerns
SMART	Offered reform platform at SEIU convention; created reform caucus Created candidate slates to run in opposition to leadership Contested SEIU elections Elected 10% of its candidates to leadership positions
Chicago Glassworkers	Held direct election for strike activities Held direct election to accept or reject owner severance proposal
UNITE HERE	Continued to support direct democracy in union decision making Explicitly gave support to democratic principles on website
MFD	Broke long-time tradition by opposing president in UMW election Created candidate slates to run in opposition to leadership Contested UMW elections Changed leadership via electoral process
Progressive movement NYPPU Active Painters Club	Fought election fixing Created a progressive caucus with slate of candidates Contested NYPPU elections Changed leadership via electoral process

Table 3
Militancy as a Strategy for Revitalization

Revitalization Effort	Militancy
CORE	Focus on societal concerns: equity in public education, protested school closings, rallied to support other unions Improved delegate roles: increased numbers of delegates, focus on mobilization and organizing, provided delegate training Communicated plans clearly and provided specific rationale for each action Increased direct action: red clothing days, proposed job actions, picketed specific administrators Members took actions before being elected without approval of CTU leadership Filed EEOC complaints against school board which targeted school closings in African-American neighborhoods Improved steward roles: provided steward training, focus on shop floor organizing Circulated petitions aimed at leadership Initiated contract campaign Rallied in cities immediately before leaders exchanged contract proposals with management 1997 strike
RMT	Broke ties with concessionist Labour Party More than 50 direct actions between 2000 and 2008 18 strikes between 2000 and 2008 Were willing to shut-down public transportation in London
Fight Back	Improved steward roles by increasing the use of grievance process Emphasized off the job organizing efforts Used the media to publicize reform efforts Filed legal challenges to leadership actions Lost militancy focus by over emphasizing election of reform candidates
Local 2, HERE	Reconnected union activities with broader social movement (Civil Rights Movement) Improved steward roles: provided steward training, increased numbers, Accepted help and tutelage from outside fellow travelers (Sirabella) Inside-out approach to organizing; organized at all internal and external union functions Initiated corporate campaign against Yale University: encouraged turning down of honorary degrees, picketed commencement Rallied and demonstrated at annual HERE meetings Took direct action against Yale trustees: picketing and withdrawal of investments form trustee owned bank Strike at Yale University
SMART	Initiated "One person, One vote" campaign to build bottom-up decision making
Chicago Glassworkers	Called for meeting with owners and Congressional representatives; owners no-showed Enlisted support from local politicians Used news media to publicize actions of owners and owners' bank by holding press conferences Held rally in downtown Chicago with support from community and broader labor movement Held six day sit-down strike at closing factory Obtained a reasonable severance package from owners
UNITE HERE	Comprehensive corporate campaigns, e.g. San Antonio Hyatt
MFD	Used legal actions and criminal prosecutions to oust corrupt leaders Strikes and walk-offs against the union to protest poor work conditions Sought and received help from Department of Labor
Progressive movement NYPPU Active Painters Club	Used news media to generate public support and create political pressure Enlisted and received help from special government prosecutors using new racketeering laws Used legal challenges to assert right to dissent Used legal actions to challenge actions of entrenched leadership

Table 4
Oligarchic Response to Revitalization Efforts

Revitalization Effort	Oligarchic Response
CORE	Maintained status quo, top-down leadership during prior reform efforts Continued cooperation with management Blocked votes on issues brought forward by CORE Distributed flyers stating "Stop Mob Action, Stop radical CORE" Accused CORE of running for office to be able to give union money to community groups Tried to disallow campaigning in the individual schools.
TDU	Used union constitution to stifle reform candidates' electoral success
RMT	Resistance came from external sources, the Labour Party and management Leadership supported rather than resisted democracy and militancy
Fight Back	Offered ameliorative measures to members Used divide and conquer tactic to play reform groups off of each other
HERE - Local 2	Disrupted Executive Board meetings with walk-outs and back turning when Sirabella spoke Used divide and conquer tactic to turn other local reform groups against each other Resistance from leaders at other locals
SMART	Legal harassment of reformers Continued top-down leadership and secret negotiations with management Threatened reformers with sanctions
Chicago Glassworkers	Leadership supported rather than resisted democracy and militancy
UNITE-HERE	Union president refused to take advice or involve membership in any union decision making Leadership developed formal plan to break apart UNITE-HERE mergers Used blogs and web-sites to attack reform leaders Slanderously accused reform leaders of misuse of union funds Used legal actions to harass reform leaders
MFD	Maximum response; Union President Boyle ordered murder of reform leader's family for which he was convicted Used ballot tampering to fix elections Forced union employees to make donations to leadership election campaigns Used divide and conquer tactics to turn retired union member against active members by increasing pension payments Used "union scabs" to limit efficacy of local mine walk-offs and strikes Used political action at every level to attack reform effort Used legal actions to harass reform leaders Held union meetings from which reformers were excluded Threatened to expel reformers Coopted progressive leaders with power and privilege Adopted election strategies to insure bureaucratic domination Used patronage coupled with "show of hands" elections to control members
Progressive movement NYPPU Active Painters Club	Used union disciplinary proceedings to deprive dissidents of work Coerced neighboring unions to make wage and working condition sections by using "union scabs" Made repeated concessions to management in order to prop up entrenched leaders positions Refused to negotiate new contracts forcing reformers to work without a contract Used divide and conquer tactics to turn painters against paperhangers Used divide and conquer tactics to turn retired union member against active members by increasing pension payments

TABLE 5
Outcomes of Revitalization Efforts with DMRev Explanations

Revitalization Effort	Revitalization Outcome	DMRev Explanation of Outcome
CORE	Election wins and gradual replacement of leadership Transformation from servicing model to member-driven model Prevented many school closings Won strikes Resisted concession demands from school board regarding work conditions and changes to compensation plan Achieved positive contract with wage increases, no increase in member-paid health care, just cause provision, and mediation and arbitration in cases of member discipline	Sustained democratic effort Broad-based member involvement from rank and file Willingness to engage in militant tactics
TDU	Achieved election successes in 1997 campaign Teamster oligarchy took back leadership in subsequent elections TDU continues to struggle to revitalize Teamsters and movement continues despite setbacks	Effective tactics from oligarchy reduced overall union revitalization Oligarchies are resilient
RMT	Membership grew contrary to trend during years 2000-2007 Strikes and other direct actions protected member interests Labour party became more responsive to union interests and no longer takes union support for granted	Willingness to advocate for members results in organizing successes Willingness to engage in militant action predicts continued union relevancy
Fight Back	Largely unsuccessful in changing union leadership Movment still exists and now focuses on fair wages for workers in fast food industry	Election efforts as main focus in ineffective Oligarchies are resilient
SMART	Successful at stopping by-law change that would have merged away local Effort continues; final results to be determined	Sustained effort is needed to affect revitalization
Chicago Glassworkers	Sit-down strike worked to mobilize community support Workers purchase the factory that was closing Union protected member jobs and prevented permanent closing of factory	Willingess to engage in militancy produces wins for members Effiective militancy motivates rather than alienates community stakeholders Militancy produces broad-based solidarity

Revitalization Effort	Revitalization Outcome	DMRev Explanation of Outcome
UNITE HERE	Reformers win elections Corporate campaigns and direct actions continue Militant actions such as boycotts continue to be effective strategies	Militant and democratic efforts are reinforcing Resistant oligarchies can be defeated with sustained effective leadership Militant action connected to broader social movements are effective
MFD	Corrupt leadership was ousted and replaced with reformers Corrupt oligarchs were convicted of crimes ranging from murder to fraud	Appropriate use of the legal system can be an effective form of militancy Overreach by resistant oligarchs can aid revitalization efforts
Progressive movement NYPPU Active Painters Club	Progressives successfully oust corrupt leaders Union has competive elections from 1935-51 Progressives become a new oligarchy during 1950's and 1960's Legal challenges were successful stopping illegal election procedures and member harassment Victories on the East Coast led to reform in California New York chapter was effectively takent over by Lucchese family in the 1960's	Appropriate use of the legal system can be an effective form of militancy Overreach by resistant oligarchs can aid revitalization efforts New progressive leaders can become new oligarchies Revitalization must be a continuous process to succeed over long term Failed revitalization can kill a union

Greve, Palmer and Pozner (2010) also note that research on the consequences of misconduct for individual organizational members is limited, particularly below the officer and director level. They specifically note that "more work also needs to be done on how organizational misconduct affects organizational members below the top management level" (Greve, Palmer, & Pozner, 2010, p. 91). And they point to the substantial variance in who does or does not get punished as an opportunity for valuable research insights.

On the other hand, there are reasons to doubt this baseline expectation for financial services professionals. We have seen complaints in recent business press post-2008 financial crisis, where for all the appearance of rotten behavior, there is a concern that individuals who are caught cheating their clients are not being punished. That is, in the case of misconduct on Wall Street specifically, there has been a groundswell of concern that the consequences are mild at best. While the U.S. government has extracted settlements and fines from financial firms, the amounts are seen as a slap on the wrist, dwarfed by the overall size of the firms/banks' profits. Furthermore, few individuals at the implicated firms have been penalized, either monetarily or via criminal prosecutions (Frontline 2014), raising concerns that there are no consequences for individuals—punishment is borne only by shareholders (Rushton 2014).

In fact, recent work by Roulet (2014) offers interesting theory and evidence suggesting that behavior which is criticized by the society at large might be rewarded by a specific industry. In particular, he finds that firms which are more criticized by the press tend to get more business in investment banking. This finding suggests that we should not expect negative consequences to misconduct for stockbrokers if the firms in the securities industry on Wall Street do not negatively stigmatize those individuals and merely view misconduct as a sign of aggressiveness.

These contradictory arguments and evidence, then, portrays an open question when it comes to the consequences of misconduct for individuals on Wall Street—particularly for those below the top management level. To make progress on this opportunity, we investigate the career consequences of one form of Wall Street misconduct: stockbrokers cheating their customers by generating higher fees through conducting unnecessary, unsuitable, or unauthorized transactions. Being caught cheating customers may damage the reputation of both the stockbroker and her employer, which could lead to loss of current position and adverse future labor market outcomes. But it could alternatively be perceived by current and potential employers in a positive light—a sign of aggressiveness—or at least a neutral light—a cost of doing business or an unlucky experience with a disgruntled client. Our primary question, then, is whether visible instances of misconduct have an impact on stockbroker careers. In particular, are they associated with higher or lower likelihood of exiting the profession and/or of being able to leave one's current employer? Exiting the industry and not being able to leave one's current employer are considered unfavorable outcomes for individuals in the securities industry where generally high mobility is expected and is associated with higher pay.

We also address Greve, Palmer and Pozner's (2010) question about sources of variance in the consequences of misconduct. In this respect, Arnold and Hagen (1992), for instance, show that client complaints against lawyers are more likely to be prosecuted the less experienced the lawyer is. This suggests misconduct may be a stronger signal to the market the less experienced the stockbroker is. Our second question, then, is whether a stockbroker's tenure moderates the impact of misconduct on the likelihood of exiting the industry or changing current employer.

To empirically examine our research questions, we draw on records of the Financial Industry Regulatory Authority (FINRA), the professional association and regulatory body for th U.S. securities industry. FINRA maintains records of every registered securities stockbroker.These records include employment history and any involvement in formal disputes with customers. We measure misconduct as disputes with customers which result in settlements or stockbrokers (and/or their employers) making restitution payments to customers.

Our sample includes 4,810 stockbrokers randomly selected from FINRA's population of 1.3 million stockbrokers. The resulting panel runs yearly from 1980 to 2013 which includes employment spells at 1,940 brokerage firms.

Using robust linear probability models and accounting for individual, firm, and time unobserved heterogeneity, we find that stockbrokers with recent misconduct suffer negative labor market consequences. Particularly, stockbrokers who experience settlements or restitution payments are 3.7% more likely to exit the industry and 15.4% less likely to be able to change employers over the next three years than those without such judgments. One interpretation of these results—which we are probing in more detail—is that stockbrokers who are caught cheating are disadvantaged in a way that they are driven from the profession and in a way that they cannot circulate to other firms. We also find that tenure does appear to moderate the effect of misconduct—such that higher tenure dampens the negative consequences of misconduct.

Future analysis will help establish more robust interpretations to build our theories of organizational misconduct as well as to inform regulatory policy in the securities industry.

We next provide a theoretical background for our investigation, describe the setting of our empirical study in more detail, provide details on our data and estimation model, present the results, and finally discuss our results and their implications.

THEORY

To theorize about the career consequences of misconduct on Wall Street, we draw from two sets of literatures which seem to offer contradictory insights—the literature on organizational misconduct and the literature on institutional logics. On the one hand, the longstanding arguments in the organizational misconduct literature seem to suggest that organizations and individuals who engage in misconduct will be penalized in two ways upon getting caught. First, they suffer an official monetary or symbolic penalty, imposed to them by a "social control agent" such as the government or a regulatory body (Greve, Palmer, & Pozner, 2010). Second, they suffer impaired future prospects, either in the form of withdrawal of business partners for organizations or limited labor market opportunities for individuals (Greve, Palmer, & Pozner, 2010). Recent empirical studies support this expectation in the way they find that officers and directors of firms implicated in accounting fraud suffer loss of positions with the focal firm and diminished subsequent job opportunities (Pozner 2008; Arthaud-Day & Certo 2006; Srinivasan 2005).

While the former punishment in the form of official penalties is of interest to legal scholars, the latter punishment in the form of limited labor market opportunities is of significant interest to scholars in organizational studies. In this respect, these scholars have proposed various theoretical mechanisms to explain the negative career consequences of misconduct. In one line of reasoning, Lorsch and MacIver (1989), for example, argue that misconduct signals to the market certain inadequacies, including unfavorable performance and quality, which will then limit future labor market opportunities for the individuals involved. In another line of reasoning, Pozner (2008), for instance, argues that to the extent to which misconduct represents deviation from accepted rules, regulations, and norms in general, it comes with reputational damage and negative stigma. The resulting stigma in turn reduces the social acceptability of those who are involved with misconduct (Carter & Feld, 2004; Kurzban & Leary, 2001) in a way that would limit their subsequent career opportunities, as others seek to dissociate themselves to lessen the threat to their identities and image (Pozner 2008). This line of reasoning further suggests that the more controllable is the deviation from the acceptable norms, the greater will be the extent to which an individual faces stigmatization (Goffman, 1986). That is to say, if the market perceives an individual to be in control of the act of misconduct, the greater will be the extent to which the market would seek to dissociate.

Taken together, these arguments seem to suggest that stockbrokers who are caught cheating their clients suffer negative consequences in two specific ways career-wise. First, they are more likely to exit the industry because the perceived inadequacies in their performance will lessen their market value or simply because they seek to "avoid difficult interactions with the untainted" (Pozner, 2008, p.145). Second, they are less likely able to change employers because other brokerage firms do not wish to associate—particularly because stockbrokers have high level of discretion/control in what they do and therefore their act of misconduct will be of a greater negative signal. In this respect then, we hypothesize that stockbrokers' visible instances of misconduct lead to negative career consequences:

- *Hypothesis 1a:* stockbrokers' visible instances of misconduct are associated with higher likelihood of exiting the profession.
- *Hypothesis 1b:* stockbrokers' visible instances of misconduct are associated with lower likelihood of being able to leave current employer

These arguments can also inform Greve, Palmer and Pozner's (2010) call for examining the sources of variance in the consequences of misconduct. In particular, these arguments seem to further suggest that the negative consequences of visible misconduct are weakened for those stockbrokers with higher tenure for two reasons. First, misconduct may be a weaker signal of inadequacies to the market the more experienced the stockbroker is since the market has more historical information on the performance and qualities of a more experienced individual to go by. Second, in a similar fashion, misconduct may be a weaker stigmatizing signal to the market for more experienced stockbrokers. In this case, in the aftermath of misconduct, others might think that such experienced perpetrators have been around long enough to know better, so there must have been something else that facilitated misconduct above and beyond their control. Arnold and Hagen's (1992) finding provide some support for these arguments as they show that client complaints against lawyers are more likely to be prosecuted the less experienced the lawyer is. In light of this, we hypothesize that higher tenure dampens the negative career consequences of stockbrokers' visible instances of misconduct:

- *Hypothesis 2a:* higher tenure weakens the positive relationship between stockbrokers' visible instances of misconduct and likelihood of exiting the profession.
- *Hypothesis 2b:* higher tenure weakens the negative relationship between stockbrokers' visible instances of misconduct and likelihood of being able to leave current employer.

On the other hand, the literature on institutional logics provides reasons to doubt the baseline expectation around the negative consequences of misconduct for financial services professionals on Wall Street. In this respect, recent research suggests that certain behavior in an industry which is criticized by the society at large might be rewarded by that industry itself. In doing so, for example, Roulet (2014) notes that "if loyalty to resistant logics is valued enough by crucial groups of stakeholders, it might be better for an actor to preserve the vilified logics rather than change" (Roulet, 2014, p. 26). He in fact finds that firms which are more criticized by the press for their societally perceived questionable behavior tend to get more business in investment banking. At the core of it, the argument is that when there is conflict between behavioral norms that an actor can adapt (e.g., engage in misconduct or not), the actor will benefit most from adapting to the norm that is local to them as opposed to the norm that is distant but is perhaps more universal (i.e., being loyal for better evaluation by peers).

These arguments seem to suggest that we should not expect negative but rather expect positive career consequences to misconduct for individuals in the securities industry on Wall Street where being caught cheating customers could be perceived by current and potential employers in a positive light—a sign of aggressiveness—or at least a neutral light—a cost of doing business or an unlucky experience with a disgruntled client. That is to say misconduct does not disadvantage stockbrokers to drive them from the profession, and that they are able to circulate to other firms. In this context, we therefore expect and hypothesize that stockbrokers' visible instances of misconduct lead to positive career consequences:

- *Hypothesis 1a*:* stockbrokers' visible instances of misconduct are associated with lower likelihood of exiting the profession.
- *Hypothesis 1b*:* stockbrokers' visible instances of misconduct are associated with higher likelihood of being able to leave current employer.

As for Greve, Palmer and Pozner's (2010) question about sources of variance in the consequences of misconduct, these arguments seem to further suggest that the positive consequences of visible misconduct are weakened for those stockbrokers with higher tenure. That is to say, misconduct early in the career will provide a greater signal of aggressiveness and loyalty to the local norms and ultimately will enhance future labor market opportunities, whereas misconduct later in the career will provide a lesser signal of aggressiveness and will raise doubt on the loyalty of the individual involved to the local norms (i.e., it will be too late to signal one's aggressiveness/loyalty later during the career to the industry). We therefore expect that higher tenure dampens the positive career consequences of stockbrokers' visible instances of misconduct:

- *Hypothesis 2a*:* higher tenure weakens the negative relationship between stockbrokers' visible instances of misconduct and likelihood of exiting the profession.

- *Hypothesis 2b*:* higher tenure weakens the positive relationship between stockbrokers' visible instances of misconduct and likelihood of being able to leave current employer.

These contradictory theoretical arguments raise an open question when it comes to the consequences of misconduct for individuals on Wall Street—particularly for those below the top management level—which set the stage for an empirical investigation.

EMPIRICAL SETTING

To empirically make progress on this opportunity, we investigate the career consequences of one form of Wall Street misconduct, namely stockbrokers cheating their customers by generating higher fees through conducting unnecessary, unsuitable, or unauthorized transactions, in the context of the U.S. securities industry.

We chose the U.S. securities industry as the setting for our empirical analysis because it satisfies several characteristics that facilitate the examination of our research questions: well- defined misconduct, relatively cheap mechanisms by which to seek visible adjudication of alleged misconduct, archives of individuals' employment history and records of misconduct, and relatively high mobility across employers.

At its core, the securities industry consists of firms that buy and sell financial securities on behalf of clients. This includes not only buying and selling existing securities, but also underwriting new securities issues; hence, the industry includes both stockbrokerages and investment banks. The boundaries of the industry are reasonably well-defined in the U.S. because securities trading is regulated under the provisions of the Securities Exchange Act of 1934. Any company that trades securities for its own account or on behalf of clients is required to register as a "broker/dealer" with the Securities and Exchange Commission (SEC) and with one of the industry's self-regulatory organizations (SROs), either FINRA or a specific stock exchange.

Employees who act as agents of broker/dealer firms (i.e., stockbrokers) must also be registered with the SEC and one of the SROs. Hence, they are often referred to as "registered representatives" (RRs). Registration as a stockbroker requires passing an exam to establish knowledge of financial securities, securities order processing, and ethical responsibilities to clients and for acceptable conduct.

As part of its mandate to regulate the licensing and professional behavior of securities stockbrokers, FINRA maintains a database of every person who is or has been registered as a securities broker, including their employment history within the securities industry and any involvement in formal customer disputes that entered the mandatory arbitration process and/or disciplinary actions by regulators. This database is publicly available, in order to allow investors to check the licensing, training, and dispute history of a potential stockbroker. In a similar way, the employers review these records when they are recruiting.

For a given stockbroker, the FINRA database includes information on who the stockbroker has been employed by (as a stockbroker) and for how long. It also includes information on whether the stockbroker has been involved in any customer disputes or regulatory actions, and what the outcomes of such disputes or actions have been.

Within the U.S. securities industry, stockbrokers' actions are governed by a set of conduct rules maintained and enforced by the SROs (principally, FINRA). These rules establish a range of ways in which stockbrokers can be responsible for failing to protect clients' interests, either through fraud or negligence (Astarita, 2008). The most common bases for disputes between customers and their stockbrokers include customers' claims of: churning, in which stockbrokers transact securities on behalf of clients solely for the purpose of charging commissions; unauthorized trading, in which stockbrokers buy or sell securities without the client's knowledge or approval; unsuitability, in which stockbrokers recommend securities that are not appropriate for the client's age or stated investment objectives; misrepresentation, in which a stockbroker fails to disclose important facts about or even misrepresents the nature of an investment; and negligence, in which a stockbroker has simply "failed to use reasonable diligence in the handling of the affairs of the customer" (Astarita, 2008).

Remedies for alleged violations of these conduct rules may be pursued in two ways: through private action by customers via a mandatory arbitration process or through public investigation and sanction by the regulator, FINRA.

Since 1989, standard contracts between customers and their stockbrokers require that disputes be resolved through mandatory binding arbitration rather than through lawsuits in the courts (Choi & Eisenberg, 2010; Choi, Fisch, & Pritchard, 2010). In arbitration, both sides represent their case to a panel of three arbitrators. The panel of arbitrators includes two public arbitrators and one industry arbitrator, where public arbitrators have minimal ties to the securities industry (and are predominantly lawyers) and are intended to bring a neutral perspective, while industry arbitrators are securities industry participants (including stockbrokers or lawyers who also work with securities firms) and are intended to bring expertise (Choi & Eisenberg, 2010; Choi, Fisch, & Pritchard, 2010).

While the decisions of arbitrator panels are likely imperfect, they represent the judgment of a panel of experts as to whether a brokerage firm and/or an individual stockbroker treated a customer in contravention of the profession's conduct code and thus seem a credible signal of whether misconduct occurred. Furthermore, this process is easier and less expensive to initiate than court-based private action. This suggests that customers likely pursue more cases than would be the case in many other settings in which the process is court-based. This then partially mitigates the gap, endemic to misconduct research (e.g., Krishnan & Kozhikode, 2014), that exists between actual versus observed misconduct.

According to Section 15A of the Securities Exchange Act of 1934 and FINRA Rule 8310, FINRA can impose a variety of sanctions on stockbrokers and securities firms that are found guilty of an infraction, including limitation (where a respondent's business activities are limited or modified), fine, censure, suspension (where a respondent's business activities are suspended for a specific period of time or until certain act is performed), and bar/expulsion (where a respondent stockbroker or firm is barred from the securities industry).

DATA, MEASURES, AND MODELS

This section presents more detail on our data, our three different but related measurements of organizational misconduct, and the econometric models we used to estimate our effects of interest.

Data

From FINRA records, we drew a random sample of 4,810 individuals from the population of the 1,301,584 people who were registered with FINRA as a securities broker in the U.S. during 1980-2013, including their employment spells at 1,940 brokerage firms.

We then collected the selected stockbrokers' complete work histories including instances of misconduct. We create a panel dataset from 1980 to 2013. The FINRA data identifies the dates of employment as a registered representative at any licensed stockbroker/dealer firm; the time when any customer disputes were filed and resolved; the manner in which those disputes were resolved (dismissal, settlement, or monetary judgment against the stockbroker); and the time that any regulatory actions were announced.

Our sample is unique because individual stockbrokers and their employers are identified and followed over time and the employment relationship between a stockbroker and his/her employer is continuously monitored.

Measures

As we discussed earlier, stockbrokers can cheat their clients by fraud or negligence. There are two ways that misconduct can be investigated and enforced. The first way is through formal complaints by clients which can either result in restitution payments after an arbitration hearing (if not dismissed) or result in a settlement. That is, client disputes might result in some kind of payment if not dismissed. The second way is through regulatory investigation which can result in limitation of activities, censure, suspension, and bar. We summarize these processes in Figure 1.

FIGURE 1
Misconduct

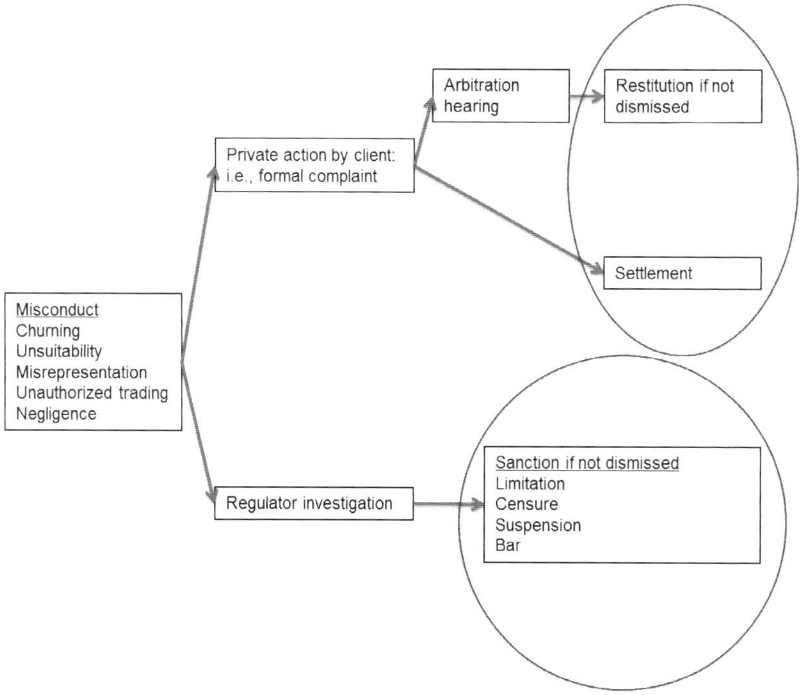

As shown in the figure with red circles, our measurement of organizational misconduct is two-fold: (1) whether or not there are disputes with customers which result in settlements or stockbrokers (and/or their employers) making restitution payments to customers; and (2) whether or not there are regulatory actions against a stockbroker—in three years prior to any given year for each individual.

We adopt a 3-year perspective in measuring misconduct to address a potential concern about reverse causality where one could argue that perhaps people first form intentions—e.g., "I'm going to leave this job or the profession soon"—then act accordingly—e.g., "since I'm going to leave, I can throw caution and cheat to make money without regard for future opportunities". We also measure misconduct as a dichotomous variable in this study to isolate the qualitative effect of misconduct. But perhaps in our future work we could explore whether there is a bigger effect for those for whom it is a second or third or n'th offense—i.e., we could address the question: do repeat offenders/offenses send a stronger signal to the market?

We also measure two specific career outcomes. Exit is set to 1 for an individual in the year beyond which we do not observe that individual in our dataset, and is set to 0 for that individual prior to that year. Employer change is set to 1 for an individual in every year when she moves to a new employer, and is set to 0 for that individual in other years. While there is a caveat in using these outcomes where it is not clear why individuals exit and whether employer change is categorically favorable or unfavorable, exiting the industry and not being able to leave one's current employer are generally considered unfavorable outcomes for individuals in the securities industry where generally high mobility is expected and is associated with higher pay. We are also collecting information on firm size and status to be able to determine if a stockbroker is perhaps "moving up or down" as they change employers.

We measure firm tenure based on the number of years an individual was employed with a firm and industry tenure based on the number of years an individual was employed in the securities industry.

Estimation Model

To test our hypotheses, we use linear probability models with robust standard errors. We do so because (a) for large number of observations it is a relatively close approximation of logistic regression which we had to use otherwise and (b) it is unbiased and does not suffer incidental parameter problem which is common for logistic models with too many fixed effects (Bennett, Pierce, Snyder, & Toffel, 2013). We also account for individual, firm, and time unobserved heterogeneity (Abowd & Kramarz, 1999a; Abowd & Kramarz, 1999b; Abowd, Kramarz, & Woodcock, 2008; Woodcock, 2011).

To do so, we estimate Equation 1:

$$y_{it} = \theta_i + \psi_{J(i,t)} + T_t + x_{it}\beta + x_{it}te_{it}\beta' + \varepsilon_{it}$$

where the dependent variable is exit/change in year t for individual i (0 or 1), the function J(i,t) indicates the employer of stockbroker i at time t, the first component is the stockbroker fixed effects, the second component is the firm fixed effects, the third component is the year fixed effects, the fourth component is organizational misconduct (restitution or settlement in three years prior to year t for individual i), the fifth component is the interaction of misconduct and firm/industry tenure (number of years), and the last component is the statistical residual, orthogonal to all other effects in the model.

RESULTS

Basic Descriptive Statistics

Table 1 presents basic statistics of our variables in our sample. This table shows that our simple random panel consists of 4,810 stockbrokers (from which 2,507 move at least once during their career) and 1,940 firms in which these stockbrokers were employed sometime in their career during 1980-2013.

This table also shows that 1.33% of the stockbrokers experience some kind of payment in 3-year periods. The average firm tenure is 5.5 years and the average industry tenure is 9.9 years. On average, 9.5% of the stockbrokers exit the industry every year while 21.5% of the stockbrokers change employers each year.

TABLE 1
Basic Descriptive Statistics

# persons	4810
# firms	1940
# mover persons	2507
Year	1980-2013
Average recent misconduct (payment)	1.33%
Average firm tenure	5.5 years
Average industry tenure	9.9 years
Average exit	9.5%
Average employer change	21.5%

Linear Probability Regression Analysis

Table 2 summarizes the main results of our regression models for when misconduct is measured as restitution payment or settlement. The table reports results from two models applied to the sample. Model 1 reports the results for exit dependent variable. Model 2 reports the results for employer change dependent variable.

As this table shows, we find that stockbrokers with recent misconduct suffer negative labor market consequences. Particularly, stockbrokers who experience settlements or restitution payments are 3.7% more likely to exit the industry and 15.4% less likely to be able to change employers over the next three years than those without such judgments. These figures are considerable given the baseline exit and employer change levels of 9.5% and 21.5% respectively. We also find that, in all cases save one, tenure does appear to moderate the effect of misconduct. In particular, we find that higher firm tenure dampens the positive relationship between misconduct and exit by 0.27%. As well, we find that higher industry tenure dampens the negative relationship between misconduct and employer change by 0.72% while higher firm tenure dampens the negative relationship between misconduct and employer change by 0.79%.

TABLE 2
Misconduct Measured as Restitution Payment or Settlement

Model	**1**		**2**	
Dependent variable	Exit		New_Spell	
Sample	Simple random		Simple random	
payment3	.0373	*	–.1541	**
	.0188		.0413	
tenure_ind	–.0028	*	–.0174	**
	.0013		.0034	
tenure_firm	.0008	*	–.0105	**
	.0003		.0007	
tenure_ind* payment3	–.0009		.0072	**
	.0013		.0022	
tenure_firm* payment3	–.0027	*	.0079	**
	.0012		.0024	
Constant	–.0173	.	.0271	.
	.		.	
Robust	Yes		Yes	
Person FE	Yes		Yes	
Firm FE	Yes		Yes	
Time FE	Yes		Yes	
# observations	50569		50569	
# persons	4810		4810	
# firms	1940		1940	
# mover persons	2507		2507	
FE F-test significant?	Yes		Yes	
r-squared	0.6726		0.2641	

Notes: Figures in smaller type are estimated robust standard errors. +p < 0.10; *p < 0.05; ** p <0.01.

We summarize these findings in Figure 2.

FIGURE 2
Misconduct Measured as Restitution Payment or Settlement

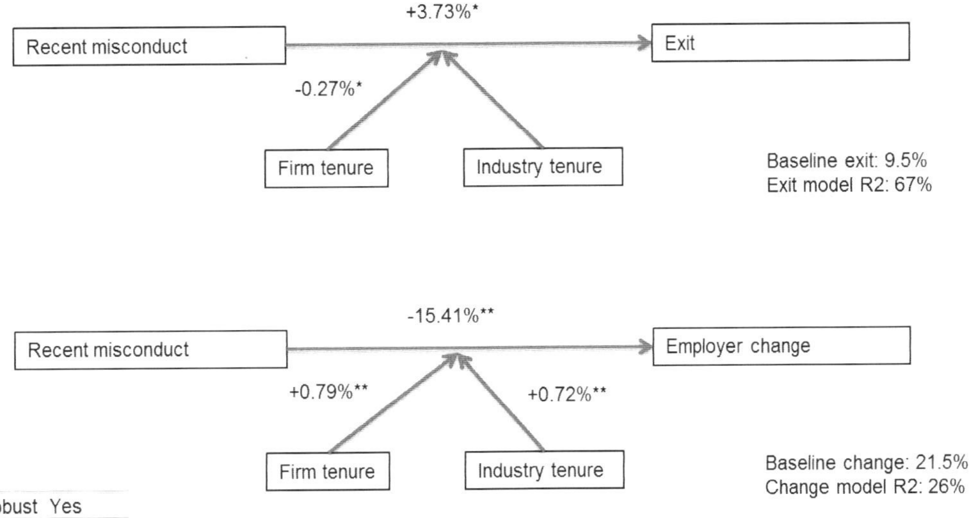

Robust	Yes
# observations	50569
FE F-test significant?	Yes
* p<0.05; ** p<0.01	

Table 3 (next page) summarizes the main results of our regression models for when misconduct is measured as regulatory sanction. The table reports results from two models applied to the sample. Model 3 reports the results for exit dependent variable. Model 4 reports the results for employer change dependent variable.

As this table shows, when we measure misconduct by whether or not a stockbroker experienced a regulatory action in a year or not, we find no significant results for our hypothesized effects. But it is unclear if this is meaningful or is a result of having smaller number of these cases.

DISCUSSION, LIMITATIONS, AND IMPLICATIONS

Using robust linear probably analyses of a random sample of stockbrokers, we address an ambiguity in our understanding of the career consequences of misconduct on Wall Street and find that stockbrokers with recent misconduct suffer negative labor market consequences, i.e., we find support for hypotheses 1a and 1b. One interpretation of these results—which we are probing in more detail—is that stockbrokers who are caught cheating are disadvantaged in a way that they are driven from the profession and in a way that they cannot circulate to other firms. We also find that tenure does appear to moderate the effect of misconduct in the way that higher tenure dampens the negative consequences of misconduct, i.e., we find support for hypotheses 2a and 2b.

There are caveats when interpreting the findings of our study though. First, at this time we do not include any observable organizational factors in our models—beyond the variables we specified earlier. Inclusion of observable organization-level characteristics such as size and status as well as additional observable individual-level characteristics such as age and asset-under- management might affect the results of our study. Second, our sample potentially suffers from survivorship bias as well as from some of the endemic issues to the organizational misconduct research, including the facts that not all misconduct is discovered/punished. Third, our estimation models might raise general concerns about endogeneity and to address that in the future we could consider doing a matched sample analysis or identifying some sample of stockbrokers that "almost" got caught cheating, as a control sample, to accurately identify the effect of getting caught itself. Our future work will refine and expand our analyses along these lines.

TABLE 3
Misconduct Measured as Regulatory Sanction

Model	**3**		**4**	
Dependent variable	Exit		New_Spell	
Sample	Simple random		Simple random	
reg3	.0256		–.0953	
	.0374		.0786	
tenure_ind	–.0028	*	–.0172	*
				*
	.0013		.0034	
tenure_firm	.0007	*	–.0103	*
				*
	.0003		.0007	
tenure_ind* reg3	.0009		.0008	
	.0026		.0047	
tenure_firm* reg3	–.0008		.0090	
	.0031		.0056	
constant	–.0166	.	.0276	.
	.		.	
Robust	Yes		Yes	
Person FE	Yes		Yes	
Firm FE	Yes		Yes	
Time FE	Yes		Yes	
# observations	50569		50569	
# persons	4810		4810	
# firms	1940		1940	
# mover persons	2507		2507	
FE F-test significant?	Yes		Yes	
R2	0.6726		0.2636	

Notes: Figures in smaller type are estimated robust standard errors.
+p < 0.10; *p < 0.05; **p < 0.01.

Notwithstanding these challenges, our study contributes to academic research on organizational misconduct. In addressing our research questions, our study validates Greve, Palmer, and Pozner's (2010) articulated baseline expectations and adds additional nuance to them—by providing evidence from below top management level and by identifying sources of variance in the consequences of misconduct. And more broadly, our study addresses calls by prominent scholars in the field of organizational misconduct by offering a systematic/objective analysis of panel data over a long period of time (rather than cross-sectional data) from actual organizations (rather than student samples) (Smith-Crowe, Tenbrunsel, Chan-Serafin, Brief, Umphress, Joseph, 2014; Craft, 2013; Kish-Gephart, Harrison, & Trevino, 2010; Tenbrunsel & Smith-Crowe, 2008). Future analysis will help establish more robust theories of organizational misconduct as well as to inform regulatory policy in the U.S. securities industry.

REFERENCES

Abowd, JM. Kramarz, F. 1999a. The Analysis of Labor Markets Using Matched Employer-Employee Data," in O. Ashenfelter and D. Card, eds. *Handbook of Labor Economics*, chapter 26, volume 3B, (Amsterdam, North Holland): 2629-2710.

Abowd, JM. Kramarz, F. 1999b. The Analysis of Labor Markets Using Matched Employer- Employee Data," *Labour Economics* 6, 53-74.

Abowd, JM., Kramarz, F., Woodcock, S. 2008. Econometric analyses of linked employer- employee data. In L. Matyas, & P. Syvestre (Eds.), *The econometrics of panel data: Fundamentals and recent developments in theory and practice* (3rd ed., pp. 729–762). Berlin: Springer.

Arnold, B. L., & Hagan, J. 1992. Careers of misconduct: The structure of prosecuted professional deviance among lawyers. *American Sociological Review*, 771-780.

Arthaud-Day, M. L., Certo, S. T., Dalton, C. M., & Dalton, D. R. 2006. A changing of the guard: Executive and director turnover following corporate financial restatements. *Academy of Management Journal*, 49(6), 1119-1136.

Astarita, Mark J. 2008. Typical customer claims in arbitration. in SECLaw.com [database online]. Available from www.seclaw.com/dispute.com.

Bennett, V., Pierce, L., Snyder, J., Toffel, M. 2013. Customer-Driven Misconduct: How Competition Corrupts Business Practices, *Management Science*, 59(8), 1725-1742.

Carter, W. C. & Feld, S. L. 2004. Principles RelatingSocial Regard to Size and Density of Personal Networks with Applications to Stigma, *Social Networks* 26(4), 323–329.

Choi, SJ., & Eisenberg, T. 2010. Punitive Damages in Securities Arbitration: An Empirical Study. *The Journal of Legal Studies.*

Choi, Stephen J., Fisch, Jill E., & Pritchard, A. C. 2010. Attorneys as Arbitrators. *Journal of Legal Studies* 39:109–58.

Craft, J. 2013. A Review of the Empirical Ethical Decision-Making Literature: 2004–2011.*Journal of Business Ethics*. Volume 117, Issue 2, pp 221-259

Frontline. 2014. Untouchables. Jan. 22. http://www.pbs.org/wgbh/pages/frontline/untouchables/ Goffman, E. 1986. *Stigma: Notes on the Management of Spoiled Identity* (Simon and Schuster, New York, N.Y).

Greve, Henrich R. Palmer, Donald. Pozner, Jo-Ellen. 2010. Organizations Gone Wild: The Causes, Processes, and Consequences of Organizational Misconduct. The *Academy of Management Annals* 4(1): 53-107.

Kish-Gephart, J. J. Harrison, D. A., & Treviño, L. K. 2010. Bad apples, bad cases, and bad barrels: Meta-analytic evidence about sources of unethical decisions at work. *Journal of Applied Psychology*. 95(1): 1-31.

Krishnan, R. & Kozhikode, R. 2014. Status and corporate illegality: Illegal loan recovery practices of commercial banks in India. Academy of Management Journal.Lipman, Abigail H. 2009. Corporate Criminal Liability. *American Criminal Law Review* 46: 359.

Kurzban, R. & Leary, M. R. 2001. Evolutionary Origins of Stigmatization: The Functions of Social Exclusion, *Psychological Bulletin* 127(2), 187–208.

Lorsch, J.W., & MacIver, E. 1989. *Pawns or potentates: The reality of America's corporate boards*. Boston: Harvard Business School Press.

Pozner, J. E. 2008. Stigma and settling up: An integrated approach to the consequences of organizational misconduct for organizational elites. *Journal of Business Ethics*, 80(1), 141-150.

Roulet, T. 2014. *There is no such thing as bad publicity: Social disapproval of investment banks as a signal of proximity to a field-level logic*. Working paper.

Rushton, Katherine. 2014. Warren Buffett calls for greater punishment of Wall Street rulebreakers. *The Telegraph*, 3 May. http://www.telegraph.co.uk/finance/businesslatestnews/10807082/Warren-Buffett-calls- for-greater-punishment-for-Wall-Street-rulebreakers.html

Smith-Crowe, K., Tenbrunsel, A. E., Chan-Serafin, S., Brief, A. P., Umphress, E. E., & Joseph, J. (forthcoming). The ethics "fix": When formal systems make a difference. *Journal of Business Ethics.*

Srinivasan, S. (2005). Consequences of financial reporting failure for outside directors: Evidence from accounting restatements and audit committee members. *Journal of Accounting Research*, 43(2), 291-334.

Tenbrunsel, A. E. & Smith-Crowe, K. (2008). Ethical decision making: Where I've been and where I're going. *Academy of Management Annals*. Vol. 2, 545-607

Woodcock, Simon D. 2011. Match Effects. Simon Fraser University. Available at http://www.sfu.ca/~swoodcoc/papers/matchfx-r12.pdf

VIII. LERA Annual Reports

VIII. LERA Annual Reports

LERA Executive Board Meeting Minutes
Monday, March 3, 2014, 3–4 p.m. Central Time
Conference Call

Call to order—The meeting was called to order at 3:05 p.m. Central Time. Present at the meeting were Bonnie Castrey, Paul Clark, Alex Colvin, Bill Dwyer, Mary MacDonald, David Lewin, and Eric Duchinsky.

Reports

Approval of the Minutes—A motion to approve the September 2013 Board Meeting Minutes passed unanimously.

Website Report—For LERA's two websites (www.leraweb.org and www.employmentpolicy.org), the host provider for both sites wants both sites upgraded to Drupal 6 (website operating system) or take the sites elsewhere. The FC recommended a plan that moves the two sites to two separate hosts (www.leraweb.org moves to MemberClicks and www.employmentpolicy.org moves to Wild Apricot). The scheduled move date is July 1st, 2014.

Fulfillment Services—The host changes initiated a review of web and member services provided by the University of Illinois Press. Savings of over $10,000 are expected by moving billing, payment processing, and member record maintenance away from UIP. Renewal notice workload will primarily move to a vendor specializing in bulk mail and printing. Record maintenance will transfer to LERA staff.

Strategic Planning for 2014—The Executive Board recommended strategic planning session requested for 2014, was delayed due to lack of funds in the 2014 budget. Strategic planning will be discussed at the Executive Board meeting in May 2014.

Financial Report—2014 budget was revisited with projection updates reflecting information collected over the past few months. Updated budget and quarterly financial performance reports will go to the Executive Board the future.

Sage Publishing Report—Jeffery Keefe, David Lewin, and Eric Duchinsky discussed the possibilities of sage publishing a peer-reviewed journal for LERA. Cynthia Nalevanko, Sage, presented information about their operation and asked for information on content. ILR Review (Cornell) recently moved to Sage. Emerald Publishing manages AILR and proposed group pricing for LERA members back in early 2013.

LAWCHA Update—The Labor and Working Class History Association (LAWCHA) has yet to reply to LERA's request for a joint meeting in 2015.

Adjournment—The meeting was adjourned at 4:15 p.m.

LERA Executive Board Meeting Minutes
Friday, May 30, 2014, 7–10 p.m. Pacific Time
Hilton Portland & Executive Tower, Portland, OR

Call to order—The meeting was called to order at 7:00 p.m. Pacific Time and finished at 10:15 p.m. Pacific Time. Present at the meeting were Ariel Avgar, Jeffery John Budd, Mary MacDonald, Bill Canak, David Lewin, Bonnie Castrey, Lisa Lynch, Alex Colvin, Ken May, Dennis Dabney, Marty Mulloy, Bill Dwyer, Susan Schurman, Janice Fine, Ami Silverman, Rebecca Givan, Chris Tilly, Marlene Heyser, and Eric Duchinsky. Not present at the meeting were Ralph Craviso, Lisa Jordan, Tazewell Hurst III, Beth Schindler, Owen Herrnstadt, and Emily Smith.

Reports

Approval of the Minutes—The motion to approve the minutes passed unanimously from the board meeting in March 2014.

LERA Annual Meeting Program Committee Report—The site for the 2016 Annual Meeting was approved: Minneapolis. Recommendations for working with ISA in a fashion similar to the 2014 Annual Meeting. There was no decision on site selection for 2017.

LERA Development Committee Report—Approved consolidation of the Development Committee and the Grants and Sponsorship Committee.

LERA Editorial Committee Report—Discussion on the strategic importance of a LERA peer-reviewed journal took place. Sage Publishing initiated a discussion with LERA about becoming the distributor for the Research Volume. The discussion triggered a consideration about the importance of LERA publishing a truly peer-reviewed academic journal. One suggestion was requesting the transfer of the AILR from Emerald Publishing to LERA. Emerald rejected this request. Another point for consideration centered beginning a new peer-reviewed journal. Things to consider included: Market saturation, initial expenses for start-up, length of time for return on start-up, expenses, length of time for achieving a respectable "impact factor" which is the most important market measurement for academic journals, will the content focus on applied, similar to the Harvard Business Review. Board Recommendation: Jeffery Keefe and David Lewin construct a position paper with recommendations working with the Editorial Committee.

Strategic Planning in 2015—Expenses for holding the strategic planning session was a main concern. Consideration for holding before the ASSA/AEA Conference in Boston, 2015 was suggested. A show of hands for those planning to attend indicated the possibility of a quorum, though concern over the balance of the Board seats was discussed.

ASSA/AEA Program Committee Report—A report on the ASSA/AEA trends reveled average attendance in 2012 was 28.8 per session, including the academic panels, workshops, and practitioner sessions. Both in 2013 and 2014, average attendance was only 7.4. Suggestions for increasing attendance includes: co-sponsoring panels with other organizations and advertising them effectively. more extensive outreach and advertising to all ASSA attendees, sponsoring a plenary that doesn't compete with many other events for attendance, sponsoring a reception, recruiting more sessions with better known scholars, scheduling the sessions likely to be the most popular at the best times, ssking people whose work is accepted for the LERA sessions to attend other LERA sessions.

Adjournment—The meeting was adjourned at 10:15 p.m.

LERA General Membership Meeting and Awards Ceremony
Saturday, May 31, 2014, 5:30 p.m Pacific Time
Pavilion Room, Hilton Portland Downtown, Portland, OR

Call to order. The meeting was called to order at 5:30 p.m. by Lisa Lynch, President.

Committee Reports

Nominating Committee Report. Martin Mulloy reported that the Nominations Committee will meet in July to develop the slate of candidates for the 2015 term. Elections are scheduled for the last week of September. Nominations for the committee's consideration should be sent to Eric Duchinsky at the LERA office.

Development and Contributions Committee Report. Marlene Heyser reported that the two fundraising committees, Development and Sponsorship/Grants merged to focus on a comprehensive plan for increasing non-dues revenue. Thanks to UAW/FORD, Kaiser Permanente, and Bloomberg/BNA supporting LERA's mission and operations.

Finance and Membership Report. Bonnie Castrey reported that in 2013 membership grew for the first time since 2006 and added 53 members over 2012. In April, the Members-only section of the LERA website went live details of content on the last page of the program guide. July 1 marks phase 2 of the website upgrade. New features and design. Please contact staff if you have any questions or ideas for content.

2014 Awards Ceremony

Best Dissertation Award Committee. Bruce Kaufman of the Thomas A. Kochan and Steven R. Sleigh Best Dissertation Award Committee awarded the Best Dissertation Award to co-winners: Alexandre Frenette, City University of New York and Barry Eidlin, University of Wisconsin-Madison. Two honorable mentions were also awarded to Maite Tapia, Michigan State University and Michael McCarthy, New York University.

LERA Awards Committee. The 2014 slate of awards were conferred. The John T. Dunlop Outstanding Scholar Awards were given to co-winners Ariel Avgar, University of Illinois at Urbana-Champaign and Matthew Bidwell, University of Pennsylvania for national research issues and the winner for international contributions was Chris Rhomberg, Fordham University. The Outstanding Practitioner Awardee was Charles Whalen, Congressional Budget Office. The Susan C. Eaton Outstanding Scholar-Practitioner Award went to Randy Eberts, W.E. Upjohn Institute for Employment Research. Academic Fellows for 2014 are David Card, Janice Bellace, and Daniel B. Cornfield. Practitioner Fellows are Martin Mulloy, Robert Taylor, and Elizabeth Shuler.

National Chapter Advisory Council Report and Awards. William Canak reported on NCAC business, and presented LERA Chapter Awards. Chapter Mentor Awardees: Chicago LERA, Gateway LERA, Central Ohio LERA, DC-LERA, Greater Houston LERA. Startup Chapter Awardee: South Florida LERA. Outstanding Chapter Awardee: Rocky Mountain LERA. Chapter Star Awardees: Alabama LERA, Oregon LERA, TERRA.

LERA Executive Board Meeting Minutes

Thursday, September 11, 2014, 3 p.m. Central Time
Conference Call

Call to order—The meeting was called to order at 3:05 p.m. Central Time. Present at the meeting were John Budd, Bonnie Castrey, Alex Colvin, Ralph Craviso (co-chair), Dennis Dabney, Eric Duchinsky, Bill Dwyer, Janice Fine, Rebecca Givan, Owen Herrnstadt, Marlene Heyser (co-chair), Tazewell Hurst III, Mary Lehman MacDonald, Lisa Lynch, Ken May, Marty Mulloy, Craig Olsen, Beth Schindler, Ami Silverman, and Chris Tilly. Not present on the call were Bill Canak, Ariel Avgar, or Emily Smith.

Reports

Approval of the Minutes—A motion to approve the June 2014 Board Meeting Minutes passed unanimously, with corrections.

Nominating Committee Report—Discussed the Nominating Committee's proposed slate of candidates. Board requested that the 2016 Nominating Committee recruit a candidate for president-elect that represents a union point of view and experience. The election slate as proposed by the nominating committee was approved unanimously.

Annual Meeting Program Committee Report—Discussed the attendee survey, financial, and word of mouth results of the Annual Meeting in Portland. Feedback was very positive and the net dollars improved over 2013. Refer to the Board report for the detailed results. One point of interest was the Thursday through Sunday schedule versus the Wednesday through Saturday schedule. The attendees were split and the Board tables the possibility of changing the scheduled days for the 2017 Annual Meeting. Confirmed the Industry Studies Association will partner with LERA in 2016 in Minneapolis, MN.

Media Award—The Board asks for additional information regarding the proposed Media Award.

Adjournment—The meeting was adjourned at 4 p.m.

LERA Executive Board Meeting Minutes
Tuesday, February 17, 2015, 1 p.m. Central Time
Conference Call

Attendees

Bill Canak
Alex Colvin
Dennis Dabney
Bill Dwyer
Janice Fine
Owen Herrnstadt
Tazewell Hurst, III
Ken May
Chris Tilly
Bonnie Castrey
Marty Mulloy
Marlene Heyser
Ralph Craviso
Eric Duchinsky
Emily Smith

Key Topics and Discussion

- Approved the minutes of the meeting was held Thursday, September 11th, 2014.
- Requested a copy of the approved meeting minutes from of Friday, May 30th, 2014 with corrections highlighted (See Attachment A). The noted correction was Owen Herrnstadt and Emily Smith were originally reported as attending the meeting when, in fact, they did not.
- Discussed the End of 2014 Budget Summary submitted by the Executive Director
    - Budget shortfall of $23,000 estimated for 2014
        - Member demographic report sent to Executive Board
    - Presented 2015 budget proposal with a shortfall of $22,350
    - Ad hoc Committee was formed to address the need for increasing new revenue for LERA. Estimated $40,000 to $50,000 will be needed annually based on current expenses. Committee to report to the Executive Board in Pittsburgh.
        - Ralph Craviso (chair)
        - Marlene Heyser
        - Janice Fine
        - Bonnie Castrey
        - David Lewin
        - Owen Herrnstadt
        - Marty Mulloy
- Presented election results for the 2015 Executive Board
- Lewin provided an update on the conference call with Ponak of the National Arbitrators Academy. Report attached as Attachment D
- Following the Executive Board's request for additional information, Lewin presented the proposal of an Annual Media Award. Report was included in the Board meeting materials for 2015-02-17 meeting.
- Due to Fueille's resignation as Secretary Treasurer and in accordance to Section II, paragraph 4 of LERA's Bylaws, "In the case of a vacancy, the President may make an interim appointment, said appointment to last until the next meeting of the Executive Board." Craig Olson, University of Illinois was put forth by Mulloy at the suggestion of Cutcher-Gershenfeld for the position of Secretary Treasurer. Nell Madigan, Associate Dean, LER School, University of Ilinois was put forth as check signer.
    - Confirmation vote was tabled with a request for more information on both candidates. See Attachment B and C. Olson will be acting Secretary Treasurer until Board votes.

- Lifetime Achievement Awards for 2015 go to:
  - Steven Greenhouse, Paul Clark nominator
  - Michael Piore, Janice Fine nominator
  - Peter Feuille, Marty Mulloy nominator
- Canak reports vitality in chapter membership
  - The merit awards process is underway.
  - NCAC workshop in Portland was very successful and will be expanded in Pittsburgh
- Annual Meeting 2015 update
  - Registration open online
  - Program released
  - Ron Bloom, Lazard, will be policy speaker
- Annual Meeting 2016 update
  - Local members not interested, at this time, in forming a chapter
  - Support from the University of MN and the local FMCS representative
  - Concerns about the hotel logistics. John Budd schedule a visit with LERA staff to address concerns.
- ASSA/AEA Meeting 2016 update
  - San Francisco
  - Castrey asks for LERA Board meeting at the ASSA/AEA meeting

The following Action Items were brought to a vote:

- Action Item: September 2014 Board Meeting Minutes: .... Approved
- Action Item: Approved 2015 budget .... Approved
- Action Item: Media Award proposal .... Approved

Action Items

- David Lewin, Bonnie Castrey, and Allen Ponak to meet in Pittsburgh on May 30th to discuss next steps.
- Duchinsky to assist Craviso with information requests for the ad hoc committee.

LERA Executive Board Meeting Minutes
Thursday, May 28, 2015, 7–10 p.m. Eastern Time
The Westin Pittsburgh, Pittsburgh, PA

Call to order—The meeting was called to order at 7:00 p.m. Pacific Time by President Marty Mulloy. Roll call was taken by Executive Director Eric Duchinsky. Board members present at the meeting were John Budd, Dennis Dabney, Rebecca Givan, Owen Herrnstadt, Beth Schindler, Alex Colvin, Bill Dwyer, Tazewell Hurst III, Ken May, Janice Fine, Mary Lehman, MacDonald, Ami Silverman, Chris Tilly, and Bill Canak. The three presidents present were Bonnie Castrey (past), Marty Mulloy (current), and Lisa Lynch (incoming). Other officers present were Craig Olson, Ariel Avgar, and Bill Canak. Committee Chairs present were Marlene Heyser, Bonnie Prouty Castrey, Jeffrey Keefe, and David Lewin. Staff present was Eric Duchinsky.

Reports

Approval of the Minutes—The motion to approve the minutes passed unanimously from the meeting on February 17, 2015 of the LERA Executive Board.

Editorial Committee Report—Ariel Avgar, Editor-in-Chief, reported on the proposed LERA 2016 Research Volume. A chapter on the impact on collective bargaining was suggested to be added by Owen Herrnstadt. Bonnie Castrey seconded the motion to approve the new volume and it was approved unanimously. Ariel Avgar will re-join the board tomorrow evening to discuss the opportunity of adding a scholarly journal to our publications.

LERA Financials Report –Financials will be overviewed in detail on Friday, May 29, 2015 at the second session LERA Executive Board meeting devoted to revenue strategies.

LERA Annual Meeting Program Committee Report–The location of next meeting will be the Hilton Anaheim in Anaheim, California. This was seconded and approved by the board. OCLERA's president elect-elect works at Disney. A tour at Disney would be an excellent special event. Disney is a major unionized work place in the area. Current Annual Meeting Update: Annual Meeting Update – attendance 341, sessions went well. Setup in Westmoreland needs to be changed for remaining days of conference. For future sessions, please ask people to commit to breakfast, lunch, dinner, etc., even if we don't sell individual tickets. When we are taking attendance counts, this information would be good to share with program committee's for planning purposes.

LERA Nominating Committee Report–Ellen Dannin is no longer at Penn State; she is currently in Ann Arbor. David Lewin is now UCLA Emeritus. The nominating committee was discussed. Diversity is key (perspective, organization, age, demographics, race, etc.). Ralph Craviso also makes the point that long-term perspective and continuity is also a key element. The board will solicit additional names and bring them to the second meeting on Friday, May 29, 2015. This topic is tabled until tomorrow evening when the board will vote on the makeup of the nominating committee. If the board does not get to this tomorrow, then the process will be to accept names from the board until the next board committee conference call.

LERA Strategic Committee Report–will be handed out at the end of the meeting.

Canadian Industrial Relations Association (CIRA) Report–Alex Colvin discusses. They propose that we might meet jointly (overlapping one day) at future meetings, in the way that we met with ISA in the past. They meet annually (50 to 100 attendees). They have overlap with our interests; they would likely wish to come to our sessions (possibility for joint registration). They are thinking of coming to the same location as us at a future meeting. They also meet around the same time as year as we do. It would be good to have the Canadian perspective, additional attendees. Lisa Lynch suggests possibly planning a joint program; if that would be the case, we would need additional time to plan the program. Peter Berg is now president elect of ISA and has mentioned that there might be some immediate plans to discuss opportunities with ISA. LAWCHA, LRAN, ISA, NAA, are all associations that have overlapping interests with LERA. LERA may be stronger if we would organize together (our meetings). It may be possible to create a council that would meet together and have joint activities. Ralph Craviso made a suggestion to be sure to capture to good ideas in an action plan with planned follow through.

ASSA-AEA Meeting Report–Conducting a board meeting at the ASSA/AEA meeting each year. We could be there to do work, follow-up from action at this meeting, recruit new members for LERA from ASSA and support the LERA sessions at the ASSA meeting. Ralph suggests that to have a second meeting might be one of the actions that could be taken by the board to follow-up. January Sixth might be the day we are discussing, all day. It might be possible to use a full discussion of present board members for say five hours, and have a portion of time (say two hours) that we include people electronically via Skype, etc. The board motioned and approved to meet on January 6, 2016 in San Francisco, CA.

LERA Development Committee–We could explore foundations that could expand the footprint of our potential contributors. Bill Dirksen and Jimmy Settles made a great presentation; we might wish to recognize them and give them a plaque. Invite the students to the award presentations. Union/Management practitioners would appreciate awards of this nature.

LERA Membership Committee–report from Jeff Keefe.

National Chapter Advisory Council Report–NCAC meeting and workshop and chapter reps meeting happening tomorrow. Baltimore has turned around as a chapter. A student chapter has formed at Rutgers and will be presenting in tomorrow's workshop. Quarterly and annual conferences that chapters have are reporting good attendance. SHRM has really vital student chapters, and it's good for LERA to replicate that student energy.

ASSA/AEA Program Committee Report–This report is accepted. Ad in Journal of Labor Economics, Society of Labor Economists; add them to our vision of LERA's big tent meeting. Come up with a format and template for action items going forward.

Adjournment—The meeting was adjourned at 10 p.m.

LERA General Membership Meeting and Awards Ceremony
Saturday, May 30, 2015, 5:30 p.m. Eastern Time
Westmoreland Room, Westin Convention Center, Pittsburgh, PA

Call to order. The meeting was called to order at 5:30 p.m. by Martin Mulloy, President.

Committee Reports

Nominating Committee Report. Martin Mulloy reported that the Nominations Committee will meet in July to develop the slate of candidates for the 2016 term. Elections are scheduled for the last week of September. Nominations for the committee's consideration should be sent to Eric Duchinsky at the LERA office.

Financial Report. Martin Mulloy reported that 2014 ended with a loss of $23,978. The Pittsburgh Conference attendees reached 350. The Executive Board created an Ad hoc Committee in February to address the strategic objects and revenue growth.

Development and Contributions Committee Report. Marlene Heyser thanked UAW/FORD, Kaiser Permanente, and Bloomberg/BNA for their significant contributions and support for LERA's mission and operations.

Finance and Membership Report. Bonnie Castrey reported that 2014 membership ended at 897, a 6% drop from 2013. In April, the Members-only section of the LERA website went live. All LERA/IRRA publications are scanned and most are posted in the searchable online database.

2015 Awards Ceremony

Best Dissertation Award Committee. Bruce Kaufman of the Thomas A. Kochan and Steven R. Sleigh Best Dissertation Award Committee awarded the Best Dissertation Award to co-winners: Tashlin Lahkani, Cornell University and Rachel Aleks, Cornell University.

LERA Awards Committee. The 2015 slate of awards were conferred by Morris Kleiner. The John T. Dunlop Outstanding Scholar Awards were given to co-winners J. Ryan Lamare, formerly of Penn State and now of University of Illinois; Adam Seth Litwin, Cornell University; and Mingwei Liu, Rutgers University. The Outstanding Practitioner Awardee was Dennis Dabney, Kaiser Permanente. The Susan C. Eaton Outstanding Scholar-Practitioner Award went to Harry Katz, Cornell University. Academic Fellows for 2015 are Greg Bamber, Monash University; Russell Lansbury, University of Sydney; and Edward Lazear, Stanford University. Practitioner Fellows are Matthew Finkin, University of Illinois at Urbana-Champaign; John S. Gaal, Carpenters' District Council of Greater St. Louis & Vicinity; and Robert Pleasure, North America's Building Trades Unions.

National Chapter Advisory Council Report and Awards. William Canak reported on NCAC business, and presented LERA Chapter Star Awards to Rocky Mountain LERA and the Tennessee Employment Relations Research Association (TERRA).

New and Other Business

The next LERA Annual Meeting will be held May 26-29, 2016 in Minneapolis, MN. After asking for any new business items and hearing none, President Martin Mulloy thanked the Executive Board and all present and handed over the gavel to Bonnie Castrey, the new LERA President. President Castrey adjourned the meeting at 6:45 p.m.

LERA Executive Board Meeting Minutes
Wednesday, September 16, 2015, 1 p.m. Easterm Time
Conference Call

Call to order—the meeting was called to order at 1 p.m. by Bonnie Castrey, President. Present at the meeting were officers President Bonnie Castrey and Secretary-Treasurer Craig Olson. Also present were board members John Budd (Program Committee Co-Chair), Alex Colvin, Bill Dwyer, Eric Fidoten, Owen Herrnstadt, Tazewell Hurst, III, Ken May, and Patrice Mareschal, and Saul Rubinstein. Unable to attend were Janice Bellace, Dennis Dabney, Rebecca Givan, Charles Jeszeck, Kris Rondeau, and Beth Schindler. LERA staff in attendance was Emily Smith.

Reports

President and Executive Committee Report—President Bonnie Castrey reports that past director Eric Duchinsky tendered his resignation effective August 15, 2015. The Executive Committee met, by conference call, reviewed the budget and financials, and appointed Emily Smith as Interim Executive Director, with a subsequent increase in pay. This motion came forward to the board as a committee action, was discussed, and ratified unanimously by the Board. The process of reviewing the budget determined that profit and loss are not meeting projections for 2015, and the officers of the organization continue to review the situation actively and will keep the Board Members updated.

Interim Director Report—Emily Smith reports that all administrative functions are complete for 2015, such as audit, insurance, and filings. The Research Volume has been mailed to members and Perspectives on Work proceeds according to schedule is to be mailed this Fall. Top priority for remainder of 2015 is to increase member dues revenue and bring back the members we have lost, if possible. We are planning five member promotions, three in September and two in October to help bring up member dues for 2015. Essentially, revenue is down in all categories, including new members, renewals, library subscribers, organizational members, and meeting income. Much of this may be linked to a new fulfillment system put in place in 2014, but additional information is needed and a full discovery will be made. Most areas of expense are up in 2015 compared to the projected budget, including Annual Meeting, Research Volume, Perspectives on Work. Steps will be taken in 2016 to bring expenses back within projections. Our short-term remedy for the losses sustained in 2015 is to maintain a full-time staff of one at the LERA office instead of two, and augment that staffing with temporary clerical help as needed, for at least the period of a year. Long-term decisions will need to be made after an inquiry is complete about the future of LERA's headquarters.

Finance Report—Secretary-Treasurer Craig Olson reports that member dues are down precipitously in 2015. This may be due to the change from fulfillment through the University of Illinois Press (UIP) to "MemberClicks" (MC). Members, library subscribers, and contributions are all down, likely connected to issues with the MC software or the user experience of it. The Annual Meeting income was also down from last year, because the meeting had 100 fewer attendees from 2014 to 2015. Almost all areas of expense are up, and need to be considered and controlled in 2016. We will keep full-time staff members to one at LERA for the next year to help with this shortfall.

Annual Meeting Profit and Loss—Secretary-Treasurer Craig Olson reports that the organization lost money on the LERA 67th Annual Meeting, May 28-31, 2015, in Pittsburgh, PA. This was carefully reviewed by staff and the causes were over-ordering of food and beverage, ordering additional breaks (food and beverage events other than meals or receptions), paying for travel and lodging of some meeting participants (which has not been done before), and paying for the registration of some meeting participants. The last three years of profit and loss for LERA Annual Meetings have been reviewed, and steps will be taken in 2016 to ensure that we have a profitable meeting. The contracts for both Minneapolis and Anaheim will be reviewed by staff and altered where needed and if possible, as they are already signed. The Minnesota venue appears to be affordable, while the Anaheim venue may not be, according to the contracts as they currently stand. Some discussion took place about the attributes of setting separate prices for registration and various events, versus having a registration price that covers all events (or most of them) at the meeting. Having a reduced registration price can be attractive to some. Having separate tickets for hospitality events can reduce inclusivity of all participants, at the conference and also increase staffing. This will continue to be evaluated.

LERA 68th Annual Meeting, May 26-29, 2016, Minneapolis Update—Bonnie Castrey reports that Emily Smith and she will be completing a site visit at the Millennium Hotel by the end of September, and reviewing the contract that we have in place. We have printed a postcard to promote the meeting and John Budd has done considerable advance promotion already, including at the Association of Labor Relations Agencies (ALRA) conference, which just took place this summer in Minneapolis. Marketing promotion begins in earnest when the program is set in January 2016.

In Minneapolis, LERA will meet jointly with the Industry Studies Association, with one day overlapping between the two, as we did in Portland 2014. Both organizations are working towards making this a collaborative day. President Bonnie Castrey and Director, Emily Smith will be meet, by conference call, with the ISA program chairs to discuss the joint meeting in Minneapolis on Friday, September 18, 2015. Additionally, Bonnie, Emily and John will meet with the ISA President and Executive Director on Wednesday, September 30, 2015, when all are in Minneapolis. Thursday, May 26 will be an overlap day and we will be discussing programming, joint plenaries, other collaborative events, hotel logistics, registration, and more.

Program Committee Arrangements Co-Chair John Budd reports that Thursday and Friday will be the key days to pull in local practitioners, something to consider as Thursday is already constrained as the overlap day between ISA and LERA. He also reports and raises concerns, that this conference is scheduled for the Memorial Day Weekend, which could prove challenging to bringing in locals who are making vacation plans following a long, potentially cold and snowy winter.

New Business

Publishing LERA Research Volume in 2016—One of the ideas put forth by the board to assist with making up for loss in 2015 was to put off the production of the LERA Research Volume in 2016. This would save the organization roughly $16,000 in expense in 2016, but we would give up any royalties for the 2016 Research Volume, which is generally between $3,000 and $6,000. We would also have to sell library subscriptions for $250 in 2016 for what would essentially be one issue of the Perspectives on Work (if purchased singly, the price is under $30), and we would have to consider if this signals poorly upon the direction of the organization. Secretary-Treasurer Craig Olson will discuss this in detail with LERA's Editor-in-Chief, Ariel Avgar, and come back to the board with advice and a recommendation.

Next Board Meeting—We are currently dealing with a number of challenges, and President Bonnie Castrey encourages all board members to attend the next board meeting in person for follow-up to these discussions, potential bylaw changes, and important issues facing the organization including strategic thinking committee findings. It will take place Tuesday, January 5, 2016, at the PARC 55 San Francisco, Market Street Room, and begin at 11am.

A call-in option will be explored for those who cannot make the meeting in person, however, **this will be a crucial meeting for this Board and all are encouraged to attend in person.**

Adjournment—The conference call meeting was adjourned at 2:10 p.m. by President Bonnie Castrey. Ken May moved the motion to the floor and Patrice Mareschal seconded it. The motion was unanimously passed by the board.

Audit Report for 2013

Labor and Employment Relations Association

Champaign, Illinois

Financial Statements

For the Years Ended
December 31, 2013 and 2012

Feller & Kuester CPAs LLP
Certified Public Accountants
806 Parkland Court, Suite 1
Champaign, IL 61821
217-351-3192

<u>TABLE OF CONTENTS</u>

Feller & Kuester CPAs LLP

Tax - Audit - Bookkeeping
806 Parkland Court, Suite #1 - Champaign, Illinois 61821
Phone - (217) 351-3192 Fax - (217) 351-4135 Email - neal@stanfellercpa.com

INDEPENDENT AUDITOR'S REPORT

To the Board of Directors of
Labor and Employment Relations Association
Champaign, Illinois 61820

We have audited the accompanying financial statements of the Labor and Employment Relations Association (a nonprofit organization) which comprise the statements of financial position as of December 31, 2013 and 2012, and the related statements of activities and cash flows for the years then ended, and the related notes to the financial statements.

Management's Responsibility for the Financial Statements
Management is responsible for the preparation and fair presentation of these financial statements in accordance with accounting principles generally accepted in the United States of America; this includes the design, implementation, and maintenance of internal control relevant to the preparation and fair presentation of financial statements that are free from material misstatement, whether due to fraud or error.

Auditor's Responsibility
Our responsibility is to express an opinion on these financial statements based on our audits. We conducted our audits in accordance with auditing standards generally accepted in the United States of America. Those standards require that we plan and perform the audit to obtain reasonable assurance about whether the financial statements are free from material misstatement.

An audit involves performing procedures to obtain audit evidence about the amounts and disclosures in the financial statements. The procedures selected depend on the auditor's judgment, including the assessment of the risks of material misstatement of the financial statements, whether due to fraud or error. In making those risk assessments, the auditor considers internal control relevant to the entity's preparation and fair presentation of the financial statements in order to design audit procedures that are appropriate in the circumstances, but not for the purposes of expressing an opinion on the effectiveness of the entity's internal control. Accordingly, we express no such opinion. An audit also includes evaluating the appropriateness of accounting policies used and the reasonableness of significant accounting estimates made by management, as well as evaluating the overall presentation of the financial statements.

We believe that the audit evidence we have obtained is sufficient and appropriate to provide a basis for our audit opinion.

Opinion
In our opinion, the financial statements referred to above present fairly, in all material respects, the financial position of the Labor and Employment Relations Association as of December 31, 2013 and 2012, and the changes in its net assets and its cash flows for the years then ended in accordance with the accounting principles generally accepted in the United States of America.

Other Matter
Our audit was conducted for the purpose of forming an opinion on the financial statements as a whole. The schedules of functional expenses on pages 16 and 17 are presented for purposes of additional analysis and are not a required part of the financial statements. Such information is the responsibility of management and was derived from and relates directly to the underlying accounting and other records used to prepare the financial statements. The information has been subjected to the auditing procedures applied in the audit of the financial statements and certain additional procedures, including comparing and reconciling such information directly to the underlying accounting and other records used to prepare the financial statements or to the financial statements themselves, and other additional procedures in accordance with auditing standards generally accepted in the United States of America. In our opinion, the information is fairly stated in all material respects in relation to the financial statements as a whole.

Feller & Kuester CPAs LLP

Feller & Kuester CPAs LLP
Champaign, IL

April 28, 2014

LABOR AND EMPLOYMENT RELATIONS ASSOCIATION
STATEMENTS OF FINANCIAL POSITION
DECEMBER 31, 2013 AND 2012

	2013	2012
ASSETS		
Current Assets		
Cash and Cash Equivalents	$ 223,184	$ 311,158
Investments	21,373	22,139
Accounts Receivable - Net	26,556	25,486
Prepaid Expenses	6,737	6,447
Inventory	6,597	6,415
Accrued Interest and Royalties	3,646	271
Total Current Assets	288,093	371,916
Property and Equipment, Net	3,112	3,846
TOTAL ASSETS	$ 291,205	$ 375,762
LIABILITIES AND NET ASSETS		
Current Liabilities		
Accounts Payable	$ 16,089	$ 15,847
Accounts Payable - UCIRHRP	4,935	4,555
Accrued Liabilities	22,059	29,993
Dues Collected in Advance	79,076	74,513
Subscriptions Collected in Advance	10,296	10,838
Deferred Chapter Dues	7,700	5,200
Other Deferred Revenue	900	-
Total Current Liabilities	141,055	140,946
Net Assets		
Temporarily Restricted		
Eaton Memorial Scholarship Receipts	34,354	37,060
Kochan-Sleigh Award Receipts	44,765	45,881
Scoville Award Receipts	7,031	7,615
Dunlop Fund	17,331	17,721
Gershenfeld Fund	17,763	12,842
Total Temporarily Restricted	121,244	121,119
Unrestricted Net Assets	28,906	113,697
Total Net Assets	150,150	234,816
TOTAL LIABILITIES AND NET ASSETS	$ 291,205	$ 375,762

See Accompanying Notes.

LABOR AND EMPLOYMENT RELATIONS ASSOCIATION
STATEMENTS OF ACTIVITIES
FOR THE YEARS ENDED DECEMBER 31, 2013 AND 2012

	2013			2012		
	Unrestricted	Temporarily Restricted	Total	Unrestricted	Temporarily Restricted	Total
Support and Revenues						
Grant Revenue						
Russell Sage Foundation Grant	$ -	$ -	$ -	$ 47,799	$ -	$ 47,799
Kauffman Foundation Grant	-	-	-	49,950	-	49,950
GE Aerospace Grant	-	-	-	7,500	-	7,500
The W.E. Upjohn Institute Grants	-	-	-	13,000	-	13,000
Total Grants	-	-	-	118,249	-	118,249
Membership Dues	142,665	-	142,665	140,867	-	140,867
Meeting Income	74,973	-	74,973	25,130	-	25,130
Organization Dues & Sponsorships	24,100	-	24,100	136,800	-	136,800
Subscriptions	20,822	-	20,822	23,758	-	23,758
Member Contributions	8,740	5,297	14,037	12,048	6,759	18,807
Chapter Fees	10,075	-	10,075	6,475	-	6,475
Royalties	6,762	-	6,762	5,854	-	5,854
Mailing List Rental	1,013	-	1,013	2,025	-	2,025
Publications	909	-	909	566	-	566
Administrative Fees	759	-	759	821	-	821
Investment Income	-	(766)	(766)	-	1,448	1,448
Interest Income	357	382	739	486	484	970
Net Assets Released from Restrictions	4,788	(4,788)	-	13,895	(13,895)	-
Total Support and Revenues	295,963	125	296,088	486,974	(5,204)	481,770
Expenses						
Program Services						
General	208,677	-	208,677	274,524	-	274,524
Meetings	52,079	-	52,079	51,751	-	51,751
Publications	44,191	-	44,191	51,595	-	51,595
Supporting Services						
Management and General	62,946	-	62,946	67,068	-	67,068
Membership Development	12,861	-	12,861	19,925	-	19,925
Total Expenses	380,754	-	380,754	464,863	-	464,863
Change in Net Assets	(84,791)	125	(84,666)	22,111	(5,204)	16,907
Net Assets, Beginning of Year	113,697	121,119	234,816	91,586	126,323	217,909
Net Assets, End of Year	$ 28,906	$ 121,244	$ 150,150	$ 113,697	$ 121,119	$ 234,816

See Accompanying Notes.

LABOR AND EMPLOYMENT RELATIONS ASSOCIATION
STATEMENTS OF CASH FLOWS
FOR THE YEARS ENDED DECEMBER 31, 2013 AND 2012

	2013	2012
CASH FLOWS FROM OPERATING ACTIVITIES		
Change in Net Assets	$ (84,666)	$ 16,907
Adjustments to Reconcile Change in Net Assets to Net Cash Provided by (Used in) Operating Activities		
Depreciation Expense	1,273	1,697
Net Unrealized (Gains) Losses on Investments	1,639	(332)
(Increase) Decrease in Operating Assets		
Accounts Receivable - Net	(1,070)	4,993
Prepaid Expenses	(290)	6,209
Inventory	(182)	940
Accrued Interest and Royalties	(3,375)	30
Increase (Decrease) in Operating Liabilities		
Accounts Payable	242	12,436
Accounts Payable - UCIRHRP	380	(488)
Accrued Liabilities	(7,934)	(7,947)
Dues Collected in Advance	4,563	(6,470)
Subscriptions Collected in Advance	(542)	(2,040)
Deferred Chapter Dues	2,500	5,200
Deferred Contributions	-	(20,675)
Deferred Grant Income	-	(90,299)
Other Deferred Revenue	900	-
Net Cash Provided by (Used in) Operating Activities	(86,562)	(79,839)
CASH FLOWS FROM INVESTMENT ACTIVITIES		
Reinvested Interest, Dividends, and Capital Gains	(873)	(1,116)
Payments for Property and Equipment	(539)	(1,599)
Net Cash Provided by (Used in) Investment Activities	(1,412)	(2,715)
CASH FLOWS FROM FINANCING ACTIVITIES		
None	-	-
Net Cash Provided by (Used in) Financing Activities	-	-
Net Increase (Decrease) in Cash and Cash Equivalents	(87,974)	(82,554)
Beginning Cash and Cash Equivalents	311,158	393,712
Ending Cash and Cash Equivalents	$ 223,184	$ 311,158

See Accompanying Notes.

LABOR AND EMPLOYMENT RELATIONS ASSOCIATION
NOTES TO FINANCIAL STATEMENTS
DECEMBER 31, 2013 AND 2012

NOTE 1 – NATURE OF ACTIVITIES AND SIGNIFICANT ACCOUNTING POLICIES

Organization and Nature of Activities

The Labor and Employment Relations Association (LERA) was founded in 1947 to encourage research in all aspects of the field of labor, employment, and the workplace. It is a nonprofit scholarly association of academic, labor, business and neutral communities committed to the full discussion and exchange of ideas between and among its broad constituencies through meetings, publications, and its various electronic listservs and websites. The LERA National Office is located in Champaign, Illinois and serves the association by planning conferences and meetings and publishing the various researches of its members. The main sources of support and revenues for LERA are grant revenue, contributions, membership dues and subscriptions, and meeting income.

Subsequent Events

Subsequent events have been evaluated through April 28, 2014, which is the date the financial statements were available to be issued.

Accrual Basis of Reporting

LERA has chosen to report on the accrual basis of accounting. Accordingly, revenue is recognized when earned and expenses are recognized when incurred in conformity with accounting principles generally accepted in the United States of America.

Estimates

The preparation of financial statements in conformity with U.S. generally accepted accounting principles requires management to make estimates and assumptions that affect certain reported amounts and disclosures. Accordingly, actual results could differ from those estimates.

Cash and Cash Equivalents

For purposes of reporting cash flows, cash and cash equivalents include all cash and highly liquid investments acquired with an original maturity date of three months or less. Since the penalties of converting certificate of deposits to cash is insignificant, all certificate of deposits have been included with cash and cash equivalents. As of December 31, 2013 and 2012, $99,871 and $98,980 of cash and cash equivalents are temporarily restricted and are subject to donor-imposed stipulations.

Investments

Investments consist of intermediate term bond index funds. All investments are held as temporarily restricted and are subject to donor-imposed stipulations.

Accounts Receivable

Accounts receivable are recorded primarily for outstanding invoices for membership dues and subscriptions. An allowance for doubtful accounts is based on an analysis of expected collection rates determined from experience. LERA had an allowance for doubtful accounts of $100 at December 31, 2013 and 2012.

Inventory

LERA's inventory of directories, research volumes, proceedings, and perspective magazines is carried at the lower of cost and market value. Cost is determined on the basis of first in – first out.

Property and Equipment

Property and equipment expenditures in excess of $500 are capitalized at cost. Donated property and equipment are capitalized at estimated cost or fair market value at the time of donation. Depreciation of the assets is computed using the straight-line method over their estimated useful lives. The range of estimated useful lives by type of asset is as follows:

Furniture and Equipment	5-7 years

Net Assets

Net assets of LERA and changes therein are classified and reported as follows:

Unrestricted Net Assets – Net assets that are not subject to donor-imposed stipulations.

Temporarily Restricted Net Assets – Net assets subject to donor-imposed stipulations that may or will be met, either by actions of LERA and/or passage of time. When a restriction expires, temporarily restricted net assets are reclassified from restrictions. Contributions received with temporarily restrictions that are met in the same reporting period are reported as unrestricted support and increase unrestricted net assets. LERA had $121,244 and $121,119 of temporarily restricted net assets as of December 31, 2013 and 2012, respectively.

Permanently Restricted Net Assets – Net assets subject to donor-imposed stipulations that they be maintained permanently by LERA. Generally, the donors of these assets permit LERA to use all or part of the income earned on any related investments for general or specific purposes. LERA had no permanently restricted net assets as of December 31, 2013 and 2012.

Grant Revenue

Grant funds received under contracts that require provision of services by LERA are recognized as revenue to the extent that expenses have been incurred for the purpose specified by the grantor. The excess of grant monies received over expenses incurred is accounted for as deferred revenue. The excess of grant expenditures over grant monies received is accounted for as grant revenue receivable to the extent that additional grant revenue has been awarded and will be paid.

Grant funds received with only general restrictions and public donations are considered to be available for unrestricted use unless specifically restricted by the donor. When such restrictions exist, the revenue is recorded as temporarily restricted until the restrictions are met.

Contributed Services and Goods

Contributed services are reported as contribution revenue and as assets or expenses only if the services create or enhance a non-financial asset (for example property and equipment) or:

- Would typically need to be purchased by LERA if the services had not been provided by contribution.
- Require specialized skills.
- Are provided by individuals with those skills (such as accounting, financial, construction, educational, electrical, legal, medical, and other services provided by accountants, investments advisers, contractors, teachers, electricians, lawyers, doctors, and other professional and craftspeople).

For the years ended December 31, 2013 and 2012, the value of contributed services meeting the defined requirements for recognition in the financial statements as outlined above were not material and have not been recorded on the financial statements. LERA does receive free office space from the University of Illinois. However, the value for the use of this office space has not been reported on the financial statements.

Income Tax Status

LERA is a not-for-profit association that is exempt from federal income tax under Section 501 (c)(3) of the Internal Revenue Code. In addition, the Internal Revenue Service has determined that LERA is not a private foundation as defined in Section 509(a)(1) and Section 170(b)(1)(A)(vi) of the Code.

LERA has evaluated its exposure resulting from uncertain income tax position and determined the exposure is not material to the financial statements. In addition, LERA is not aware of any tax position for which a significant change is reasonably possible within the next 12 months. Therefore, these financial statements do not include a liability for uncertain tax positions. Upon recognition of a liability for an uncertain tax position, LERA would recognize interest expense and penalties in operating expenses.

LERA files information tax returns in the U.S. federal jurisdiction and the state of Illinois. Its federal and Illinois information tax returns prior to fiscal year 2010 are closed. LERA does not have any tax returns currently under examination by either the Internal Revenue Service (IRS) or any U.S. state jurisdiction.

Membership Dues and Advance Subscriptions Collected

Membership dues and subscriptions are assessed and recognized as revenue based on the life of the dues or subscription.

Functional Allocation of Expenses

The costs of providing the various programs and other activities have been summarized on a functional basis in the statement of activities. Accordingly, certain costs have been allocated among the programs and supporting services benefited.

NOTE 2 – ARRANGEMENTS WITH THE UNIVERSITY OF ILLINOIS

LERA moved its offices to the University of Illinois at the end of 1999. Under an arrangement with the University, the employees of LERA are employed by the University. The employees' pension and benefits are part of the University's plans. LERA then reimburses the University quarterly for the cost of its employees.

The University of Illinois holds some cash for LERA. These "claim on cash" balances were $29,804 and $67,735 as of December 31, 2013 and 2012, respectively. These balances have been included with cash and cash equivalents.

NOTE 3 – UCIRHRP SERVICING ARRANGEMENT TO ADMINISTER BUSINESS

On January 3, 2003, LERA entered into a servicing arrangement to administer the business of the University Council of Industrial Relations and Human Resources Programs (UCIRHRP). These servicing duties include dues notification and collection, annual meeting arrangement and report preparation, and maintaining a data base and network communications for this separate organization. A separate financial statement is maintained for this company, but UCIRHRP's cash is maintained in LERA's general bank account. The amount of cash belonging to UCIRHRP in the general account is shown on LERA's balance sheet as Accounts Payable - UCIRHRP. UCIRHRP's cash balance in LERA's custody was $4,935 at December 31, 2013 and $4,555 at December 31, 2012. LERA is allowed to keep 20% of dues collected each year as an administrative fee.

NOTE 4 – CONCENTRATIONS OF CREDIT RISK

Financial instruments that potentially subject LERA to credit risk consist principally of cash and cash equivalents at financial instructions. However, management continuously monitors LERA's balances at financial institutions. Deposits held at any financial institutions did not exceed depository insurance at December 31, 2013 or 2012. The "claims on cash" held with the University of Illinois as stated in Note 2 is not insured since it is not held in a financial institution.

NOTE 5 – GRANT REVENUE

Russell Sage Foundation Grant

On September 2, 2010, LERA received notification that it was the recipient of a two-year appropriation grant for $149,469 to support the creation of the EPRN within LERA as fiscal agent for the award. The grant was received in two installments, with $72,487 received on

November 1, 2010, and the final payment of $76,982 was received on July 11, 2011. The 2010 installment was not offset by any 2010 expenses, and the entire amount was deferred to 2011. The grant money was offset with 2011 expenses totaling $101,670, which includes overhead charges allowed at 15%. The remaining $47,799 of unspent grant funds deferred to 2012 was expended by June 30, 2012 which was the end of the grant period.

Kauffman Foundation Grant

On November 22, 2011, LERA received a fifteen month appropriation grant for $49,950 to support the EPRN research within LERA. $35,000 was received on December 14, 2011, and the final payment of $14,950 was received on December 14, 2012. The entire $49,950 was expensed during the year ended December 31, 2012.

GE Aerospace Grant

LERA received funds in December of 2005 toward the next Aerospace Industry Council. As of December 31, 2011, $7,500 of these grant receipts had not yet been expensed and thus had been deferred. These funds were expensed in 2012, and therefore recorded as grant revenue during 2012.

The W.E. Upjohn Institute Grants

On January 26, 2012, LERA received grants to support the writing of working papers on "Financialization and Labor Market Outcomes" and "America's Human Capital Paradoxes" for a total of $20,000. As of December 31, 2012, $13,000 of the grant funds had been received and earned.

NOTE 6 – TEMPORARILY RESTRICTED NET ASSETS

Susan C. Eaton Scholar-Practitioner Memorial Fund

LERA set up a memorial fund in honor of an author of a 1998 "Perspectives on Work" article and active young LERA member, who died on December 30, 2003. At its June 1, 2004 meeting, LERA approved the establishment of an annual Susan C. Eaton Scholar-Practitioner award and grant to be paid to one or more qualified scholar researchers in even-numbered years or practitioners in odd-numbered years doing research in the labor and employment relations or related field. LERA's Executive Board directed that 10% of any gifts received each year are unrestricted and can be used for administrative expenses. As of December 31, 2013 and 2012, these temporary restricted funds totaled $34,354 and $37,060, respectively.

Kochan-Sleigh Best Dissertation Award Fund

On March 27, 2006, LERA set up a temporarily restricted fund with contributions from a member and matching contributions from General Electric. The fund is restricted for a minimum of fifteen years and may be used to pay for the $1,000 annual best dissertation award and plaque. After the fifteen-year period, the fund may be continued or the amount remaining in the fund may be released and become available to LERA for unrestricted purposes. LERA's Executive Board directed that 10% of any gifts received each year are unrestricted and can be used for administrative expenses. As of December 31, 2013 and 2012, these temporary restricted funds totaled $44,765 and $45,881, respectively.

Jim Scoville Award Fund

On January 9, 2009, LERA set up a temporarily restricted fund with a $10,000 contribution from the University of Minnesota's Industrial Relations Center to honor a member and retiring professor. The fund was established to pay an annual award of $500 and a plaque for best paper on international and comparative employment issues. LERA's Executive Board directed that 10% of any gifts received each year are unrestricted and can be used for administrative expenses. As of December 31, 2013 and 2012, these temporarily restricted funds totaled $7,031 and $7,615, respectively.

John T. Dunlop Public Policy Fund

On May 8, 2010, LERA established a temporarily restricted fund to subsidize a named John T. Dunlop Public Policy Session at future LERA National Policy Forums and/or at LERA annual meetings. A total of $252, $1,386, $2,858, and $21,025 of donations were collected during 2013, 2012, 2011, and 2010, respectively. LERA's Executive Board directed that 10% of any gifts received each year are unrestricted and can be used for administrative expenses. This fund will be so named for five or ten years, at which time LERA's Executive Board will review the fund to determine whether to continue or modify the fund. As of December 31, 2013 and 2012, these temporary restricted funds totaled $17,331 and $17,721, respectively.

Gladys and Walter Gershenfeld Publication Fund

On May 8, 2010, LERA established a temporarily restricted fund to honor the memory of long-time LERA members and LERA Past President Walter and Gladys Gershenfeld for the purpose of supporting LERA's electronic and print publications. A total of $5,045, $5,373, $5,917, and $5,450 of donations were collected during 2013, 2012, 2011, and 2010, respectively. The fund will be so named for ten years, at which time LERA will review the fund's purpose and uses, and whether to continue or modify the fund. LERA's Executive Board directed that 10% of any gifts received each year are unrestricted and can be used for administrative expenses. As of December 31, 2013 and 2012, these temporarily restricted funds totaled $17,763 and $12,842, respectively.

NOTE 7 – INVESTMENTS

Financial Accounting Standards Board ("FASB") Codification 820, Fair Value Measurements, establishes a framework for measuring fair value. That framework provides a fair value hierarchy that prioritizes the inputs to valuation techniques used to measure fair value. The hierarch gives the highest priority to unadjusted quoted prices in active markets for identical assets or liabilities (Level 1 measurements) and the lowest priority to unobservable inputs (Level 3 measurements). The three levels of the fair value hierarchy under FASB Codification 820 are described below:

Level 1 – Inputs to the valuation methodology are based on unadjusted quoted prices for identical assets or liabilities in active markets that LERA has the ability to access.

Level 2 – Inputs to the valuation methodology include quoted prices for similar assets or liabilities in active markets, quoted prices for identical or similar assets or liabilities in inactive markets, inputs other than quoted prices that are observable for the asset or liability, and inputs that are derived principally from or corroborated by observable market data by correlation or other means. If the asset or liability has a specified (contractual) term, the Level 2 input must be observable for substantially the full term of the asset or liability.

Level 3 – Inputs to the valuation methodology are unobservable and significant to the fair value measurement.

The asset's and liability's fair value measurement level within the fair value hierarchy is based on the lowest level of any input that is significant to the fair value measurement. Valuation techniques used need to maximize the use of observable inputs and minimize the use of unobservable inputs.

The following is a description of the valuation methodologies used for assets measured at fair value. There have been no changes in the methodologies used at December 31, 2013 and 2012.

Intermediate Term Bond Funds: Valued at fair market value.

The preceding methods described may produce a fair value calculation that may not be indicative of net realizable value or reflective of future fair values. Furthermore, although LERA believes its valuation methods are appropriate and consistent with other market participants, the use of different methodologies or assumptions to determine the fair value of certain financial instruments could result in different fair value measurements at the reporting date.

The following table sets forth by level, within the fair value hierarchy, LERA's assets at fair value as of December 31, 2013 and 2012:

Assets at Fair Value as of December 31, 2013

	Level 1	Level 2	Level 3	Total
Inter-term Bond Funds	$21,373	-	-	$21,373

Assets at Fair Value as of December 31, 2012

	Level 1	Level 2	Level 3	Total
Inter-term Bond Funds	$22,139	-	-	$22,139

LABOR AND EMPLOYMENT RELATIONS ASSOCIATION
NOTES TO FINANCIAL STATEMENTS
DECEMBER 31, 2013 AND 2012

Net investment earnings for the years ended December 31, 2013 and 2012 are summarized as follows:

	2013	2012
Dividends	$ 637	$ 678
Capital Gains Distributions	236	438
Net Unrealized Gains (Losses)	(1,639)	332
Total Investment Income	$ (766)	$ 1,448

This investment income, including unrealized gains and losses, are being reported on the statement of activity as investment income.

NOTE 8 – PROPERTY AND EQUIPMENT

As of December 31, 2013, Property and Equipment consists of:

Furniture and Equipment	$ 17,347
Less: Accumulated Depreciation	(14,235)
Property and Equipment, Net	$ 3,112

As of December 31, 2012, Property and Equipment consists of:

Furniture and Equipment	$ 20,756
Less: Accumulated Depreciation	(16,910)
Property and Equipment, Net	$ 3,846

Depreciation expense for the years ended December 31, 2013 and 2012 was $1,273 and $1,697, respectively.

NOTE 9 – ACCRUED EXPENSES

As of December 31, 2013, Accrued Expenses consist of:

Accrued Payroll	$ 5,879
Accrued Benefits/Taxes	2,118
Accrued Vacation	14,062
Total	$ 22,059

As of December 31, 2012, Accrued Expenses consist of:

Accrued Payroll	$ 8,548
Accrued Benefits/Taxes	3,402
Accrued Vacation	18,043
Total	$ 29,993

NOTE 10 – COMMITMENTS

LERA has a contract with University of Illinois Press (UIP) for UIP to provide membership management services and publication of *Perspectives on Work* (magazine) for an annual fee of $28,280. The five year agreement commenced on January 1, 2007 and expired on December 31, 2012. Although the contract expired, it was still in place as of December 31, 2013 and 2012. However, LERA will end this relationship with UIP on July 1, 2014.

On February 26, 2013, LERA signed a 12 month contract from March 1, 2013 to February 28, 2014 with a company to provide services to manage the EPRN website and provide content for the LERA website. The contract states that LERA will pay this company $2,400 per month for a total of $28,800. The Agreement automatically renews for consecutive 12 month terms, unless the Agreement is terminated by one or both of the parties as set forth herein. Beginning the second cycle and continuing until termination or renegotiation, LERA will increase the base rate every year by two percent (2%) over the previous year's base rate. For example: $2,448 per month for twelve months starting March 1, 2014; $2,497 per month for twelve months starting March 1, 2015; $2,546 per month for twelve months starting March 1, 2016; and so on.

LERA has signed contracts with hotels in St. Louis, Portland, and Pittsburgh for the 2013, 2014, and 2015 Annual Meetings. If these contracts were cancelled at December 31, 2013 and 2012, LERA would have owed $76,903 and $33,151 to the St. Louis, Portland, and Pittsburgh hotels, respectively.

NOTE 11 – PENSION PLAN AND RETIREMENT CONTRIBUTIONS

The University of Illinois offers participation in the State Universities Retirement System of Illinois (SURS), a cost-sharing multiple-employer defined-benefit pension plan with a special funding situation whereby the State makes substantially all actuarially determined required contributions on behalf of the participating employers. SURS was established July 21, 1941, to provide retirement annuities and other benefits for staff members and employers of the State universities, certain affiliated organization, and certain other State educational and scientific agencies and for survivors, dependents and other beneficiaries of such employees. SURS is considered a component unit of the State of Illinois' financial reporting entity, and is included in the State's financial reports as a pension trust fund. SURS is governed by Section 5/15, Chapter 40, of the Illinois Compiled Statutes. SURS issues a publically available financial report that includes financial statements and required supplementary information. That report may be obtained by accessing the Web site at www.SURS.org or by calling 1-800-275-7877.

Eligible employees must participate upon initial employment. Employees are ineligible to participate if (a) employed after having attained age 68; (b) employed less than 50% of full-time; or (c) employed less than full-time and attending classes with an employer.

Plan members are required to contribute 8.0% of their annual covered salary and substantially all employer contributions are made by the State on behalf of the individual employers at an actuarially determined rate. LERA's current employer rate is 11.91% of annual covered payroll. The contribution requirements of plan members and employers are established and may be amended by the Illinois General Assembly. During the years ended December 31, 2013, 2012 and 2011, LERA reimbursed the University of Illinois $16,436, $20,632, and $18,001, respectively, for the employer's payments to SURS for eligible employees.

Employers hired prior to April 1, 1986 are exempt from contributions required under the Federal Insurance Contribution Act. Employees hired after March 31, 1986 are required to contribute 1.45% of their gross salary for Medicare. LERA is required to match this contribution. As of December 31, 2013 and 2012, all employees were required to contribute 1.45% of their gross salary for Medicare.

Employees may also elect to participate in certain tax-sheltered retirement plans. These voluntary plans permit employees to designate a part of their earnings into tax-sheltered investments and thus defer federal and state income taxes on their contribution and accumulated earnings under the plans. Participation and the level of employee contributions are voluntary. The employer is not required to make contributions to these plans.

NOTE 12 – POST-EMPLOYMENT BENEFITS

In addition to providing the above pension benefits, the State provides health, dental, vision, and life insurance benefits for retirees and their dependents in a program administered by the Department of Central Management Services. Substantially all State employees become eligible for post-employment benefits if they eventually become annuitants of one of the State sponsored plans. Health, dental, and vision benefits include basic benefits for annuitants and dependents under the State's self-insurance plan and insurance contracts currently in force. All retirees, annuitants, and survivors will be charged up to two percent of their annuity value to cover the costs of the basis program of group health benefits. In addition to this percentage, some annuitants and survivors are required to make further contributions toward their health and dental benefits. Annuitants receive life insurance coverage equal to the annual salary of the last day of employment until age 60, at which time the benefit becomes $5,000. The State pays the University's portion of employer costs for the benefits provided. The total costs of the State's portion of health, dental, vision, and life insurance benefits of all members, including post-employment health, dental, vision, and life insurance benefits, is recognized as an expenditure by the State in the Illinois Comprehensive Annual Financial Report. A copy of the financial statements of the State can be obtained at www.ioc.state.il.us. The State finances the costs on a pay-as-you-go basis. The total costs incurred for health, dental, vision, and life insurance benefits are not separated by department or component unit for annuitants and their dependents nor active employees and their dependents.

LABOR AND EMPLOYMENT RELATIONS ASSOCIATION
SCHEDULE OF FUNCTIONAL EXPENSES
FOR THE YEAR ENDED DECEMBER 31, 2013

		MEETINGS					PUBLICATIONS						SUPPORTING SERVICES		
	General	Annual Conference	Spring Bd Meeting	ASSA Meeting	Winter Bd Meeting	Other Meetings	Directory	Newsletter	Annual Proceedings	Perspectives	Research Volume	Other Publications	Management & General	Membership Development	Totals
Compensation	$ 131,006	$ -	$ -	$ -	$ -	$ -	$ -	$ -	$ -	$ -	$ -	$ -	$ 14,396	$ -	$ 145,402
Payroll Taxes & Fringes	52,911	-	-	-	-	-	-	-	-	-	-	-	6,434	-	59,345
Contract Labor	24,000	-	-	-	-	-	-	-	-	-	-	-	-	-	24,000
Depreciation	-	-	-	-	-	-	-	-	-	-	-	-	1,273	-	1,273
Insurance	-	-	-	-	-	-	-	-	-	-	-	-	3,653	-	3,653
Bank Charges	-	-	-	-	-	-	-	-	-	-	-	-	1,882	-	1,882
Promotion	-	8,111	-	-	-	-	-	-	-	-	-	-	-	4,395	12,506
Fulfillment	-	-	-	-	-	-	-	-	-	-	-	-	13,180	-	13,180
Postage and Freight	-	-	-	-	-	-	-	-	-	1,102	2,221	23	281	-	3,627
Accounting/Auditing	-	-	-	-	-	-	-	-	-	-	-	-	4,800	-	4,800
Printing and Production	-	1,238	-	-	-	-	-	-	-	17,893	5,415	60	-	-	24,606
Services	-	-	-	-	-	-	-	-	1,188	4,058	8,360	-	-	-	13,606
Other Publication Costs	-	-	-	-	-	-	1,750	750	1,000	352	-	19	-	-	3,871
Meals, Receptions	-	28,967	-	-	-	-	-	-	-	-	-	-	-	-	28,967
Travel	-	6,469	-	503	-	725	-	-	-	-	-	-	-	-	7,697
Other Meeting Expenses	-	4,687	-	114	-	-	-	-	-	-	-	-	-	-	4,801
Education	-	-	-	-	-	-	-	-	-	-	-	-	-	-	-
Computer Supplies/Services	-	1,265	-	-	-	-	-	-	-	-	-	-	5,296	-	6,561
Website	-	-	-	-	-	-	-	-	-	-	-	-	9,432	-	9,432
Office Supplies	760	-	-	-	-	-	-	-	-	-	-	-	612	-	1,372
Student & Member Awards	-	-	-	-	-	-	-	-	-	-	-	-	-	6,097	6,097
Fundraising Expense	-	-	-	-	-	-	-	-	-	-	-	-	-	1,650	1,650
Duplicating Expense	-	-	-	-	-	-	-	-	-	-	-	-	1,047	-	1,047
Telephone and Fax	-	-	-	-	-	-	-	-	-	-	-	-	476	-	476
Chapter Expenses	-	-	-	-	-	-	-	-	-	-	-	-	-	292	292
Dues	-	-	-	-	-	-	-	-	-	-	-	-	65	-	65
Other Committee Expenses	-	-	-	-	-	-	-	-	-	-	-	-	-	427	427
Miscellaneous Office	-	-	-	-	-	-	-	-	-	-	-	-	119	-	119
	$ 208,677	$ 50,737	$ -	$ 617	$ -	$ 725	$ 1,750	$ 750	$ 2,188	$ 23,405	$ 15,996	$ 102	$ 62,946	$ 12,861	$ 380,754

LABOR AND EMPLOYMENT RELATIONS ASSOCIATION
SCHEDULE OF FUNCTIONAL EXPENSES
FOR THE YEAR ENDED DECEMBER 31, 2012

		MEETINGS					PUBLICATIONS						SUPPORTING SERVICES		
	General	Annual Conference	Spring Bd Meeting	ASSA Meeting	Winter Bd Meeting	Other Meetings	Directory	Newsletter	Annual Proceedings	Perspectives	Research Volume	Other Publications	Management & General	Membership Development	Totals
Compensation	$ 193,354	$ -	$ -	$ -	$ -	$ -	$ -	$ -	$ -	$ -	$ -	$ -	$ 18,239	$ -	$ 211,593
Payroll Taxes & Fringes	75,170	-	-	-	-	-	-	-	-	-	-	-	7,254	-	82,424
Contract Labor	6,000	-	-	-	-	-	-	-	-	-	-	-	-	-	6,000
Depreciation	-	-	-	-	-	-	-	-	-	-	-	-	1,697	-	1,697
Insurance	-	-	-	-	-	-	-	-	-	-	-	-	3,508	-	3,508
Bank Charges	-	-	-	-	-	-	-	-	-	-	-	-	609	-	609
Promotion	-	10,366	-	-	-	-	-	-	-	-	-	-	-	6,604	16,970
Fulfillment	-	-	-	-	-	-	-	-	-	-	-	-	13,180	-	13,180
Postage and Freight	-	561	-	-	-	-	-	-	-	1,517	2,443	-	847	-	5,368
Accounting/Auditing	-	-	-	-	-	-	-	-	-	-	-	-	4,800	-	4,800
Printing and Production	-	934	-	-	-	-	-	-	-	16,700	4,664	58	-	-	22,366
Services	-	-	-	-	-	-	-	-	-	-	-	-	-	-	-
Other Publication Costs	-	-	-	-	-	-	1,750	750	2,388	10,831	7,408	3,076	-	-	26,203
Meals, Receptions	-	27,823	1,138	-	3,526	185	-	-	-	-	-	-	-	-	32,672
Travel	-	3,025	1,714	-	-	1,741	-	-	-	-	-	-	-	-	6,480
Other Meeting Expenses	-	638	-	-	-	100	-	-	-	-	-	-	-	-	738
Education	-	-	-	-	-	-	-	-	-	-	-	-	772	-	772
Computer Supplies/Services	-	-	-	-	-	-	-	-	-	-	-	-	9,971	-	9,971
Website	-	-	-	-	-	-	-	-	-	-	-	-	-	-	-
Office Supplies	-	-	-	-	-	-	-	-	-	-	-	-	423	-	423
Student & Member Awards	-	-	-	-	-	-	-	-	-	-	-	-	-	8,449	8,449
Fundraising Expense	-	-	-	-	-	-	-	-	-	-	-	-	-	2,630	2,630
Duplicating Expense	-	-	-	-	-	-	-	-	-	-	-	-	295	-	295
Telephone and Fax	-	-	-	-	-	-	-	-	-	-	-	-	1,171	-	1,171
Chapter Expenses	-	-	-	-	-	-	-	-	-	-	-	-	-	1,280	1,280
Dues	-	-	-	-	-	-	-	-	-	-	-	-	570	-	570
Other Committee Expenses	-	-	-	-	-	-	-	-	-	-	-	-	-	962	962
Miscellaneous Office	-	-	-	-	-	-	-	-	-	-	-	-	3,732	-	3,732
	$ 274,524	$ 43,347	$ 2,852	$ -	$ 3,526	$ 2,026	$ 1,750	$ 750	$ 2,388	$ 29,048	$ 14,515	$ 3,144	$ 67,068	$ 19,925	$ 464,863

Audit Report for 2014

Labor and Employment Relations Association

Champaign, Illinois

Financial Statements

For the Years Ended
December 31, 2014 and 2013

Feller & Kuester CPAs LLP
Certified Public Accountants
806 Parkland Court, Suite 1
Champaign, IL 61821
217-351-3192

TABLE OF CONTENTS

Feller & Kuester CPAs LLP

Tax - Audit - Bookkeeping
806 Parkland Court, Suite #1 - Champaign, Illinois 61821
Phone - (217) 351-3192 Fax - (217) 351-4135 Email - neal@fellerkuester.com

INDEPENDENT AUDITOR'S REPORT

To the Board of Directors of
Labor and Employment Relations Association
Champaign, Illinois

We have audited the accompanying financial statements of the Labor and Employment Relations Association (a nonprofit organization) which comprise the statements of financial position as of December 31, 2014 and 2013, and the related statements of activities and cash flows for the years then ended, and the related notes to the financial statements.

Management's Responsibility for the Financial Statements

Management is responsible for the preparation and fair presentation of these financial statements in accordance with accounting principles generally accepted in the United States of America; this includes the design, implementation, and maintenance of internal control relevant to the preparation and fair presentation of financial statements that are free from material misstatement, whether due to fraud or error.

Auditor's Responsibility

Our responsibility is to express an opinion on these financial statements based on our audits. We conducted our audits in accordance with auditing standards generally accepted in the United States of America. Those standards require that we plan and perform the audit to obtain reasonable assurance about whether the financial statements are free from material misstatement.

An audit involves performing procedures to obtain audit evidence about the amounts and disclosures in the financial statements. The procedures selected depend on the auditor's judgment, including the assessment of the risks of material misstatement of the financial statements, whether due to fraud or error. In making those risk assessments, the auditor considers internal control relevant to the entity's preparation and fair presentation of the financial statements in order to design audit procedures that are appropriate in the circumstances, but not for the purposes of expressing an opinion on the effectiveness of the entity's internal control. Accordingly, we express no such opinion. An audit also includes evaluating the appropriateness of accounting policies used and the reasonableness of significant accounting estimates made by management, as well as evaluating the overall presentation of the financial statements.

We believe that the audit evidence we have obtained is sufficient and appropriate to provide a basis for our audit opinion.

Opinion
In our opinion, the financial statements referred to above present fairly, in all material respects, the financial position of the Labor and Employment Relations Association as of December 31, 2014 and 2013, and the changes in its net assets and its cash flows for the years then ended in accordance with the accounting principles generally accepted in the United States of America.

Other Matter
Our audit was conducted for the purpose of forming an opinion on the financial statements as a whole. The schedules of functional expenses on pages 16 and 17 are presented for purposes of additional analysis and are not a required part of the financial statements. Such information is the responsibility of management and was derived from and relates directly to the underlying accounting and other records used to prepare the financial statements. The information has been subjected to the auditing procedures applied in the audit of the financial statements and certain additional procedures, including comparing and reconciling such information directly to the underlying accounting and other records used to prepare the financial statements or to the financial statements themselves, and other additional procedures in accordance with auditing standards generally accepted in the United States of America. In our opinion, the information is fairly stated in all material respects in relation to the financial statements as a whole.

Feller & Kuester CPAs LLP

Feller & Kuester CPAs LLP
Champaign, Illinois

May 19, 2015

LABOR AND EMPLOYMENT RELATIONS ASSOCIATION
STATEMENTS OF FINANCIAL POSITION
DECEMBER 31, 2014 AND 2013

	2014	2013
ASSETS		
Current Assets		
Cash and Cash Equivalents	$ 226,823	$ 223,184
Investments	22,856	21,373
Accounts Receivable, Net	250	26,556
Prepaid Expenses	6,256	6,737
Inventory	5,537	6,597
Accrued Interest and Royalties	4,305	3,646
Total Current Assets	266,027	288,093
Property and Equipment, Net	1,587	3,112
TOTAL ASSETS	$ 267,614	$ 291,205
LIABILITIES AND NET ASSETS		
Current Liabilities		
Accounts Payable	$ 8,085	$ 16,089
Funds Held for Other Organizations	7,516	4,935
Accrued Liabilities	24,315	22,059
Dues Collected in Advance	56,757	79,076
Subscriptions Collected in Advance	7,278	10,296
Deferred Sponsorships	25,000	-
Deferred Chapter Dues	8,560	7,700
Other Deferred Revenue	1,755	900
Total Current Liabilities	139,266	141,055
Net Assets		
Temporarily Restricted		
Eaton Memorial Scholarship Receipts	31,935	34,354
Kochan-Sleigh Award Receipts	43,624	44,765
Scoville Award Receipts	7,041	7,031
Dunlop Fund	19,474	17,331
Gershenfeld Fund	23,030	17,763
Total Temporarily Restricted	125,104	121,244
Unrestricted Net Assets	3,244	28,906
Total Net Assets	128,348	150,150
TOTAL LIABILITIES AND NET ASSETS	$ 267,614	$ 291,205

See Accompanying Notes.

LABOR AND EMPLOYMENT RELATIONS ASSOCIATION
STATEMENTS OF ACTIVITIES
FOR THE YEARS ENDED DECEMBER 31, 2014 AND 2013

	2014			2013		
	Unrestricted	Temporarily Restricted	Total	Unrestricted	Temporarily Restricted	Total
Support and Revenues						
Membership Dues	137,775	-	137,775	142,665	-	142,665
Meeting Income	105,080	-	105,080	74,973	-	74,973
Organization Dues & Sponsorships	70,800	-	70,800	24,100	-	24,100
Subscriptions	20,005	-	20,005	20,822	-	20,822
Member Contributions	9,772	6,422	16,194	8,740	5,297	14,037
Royalties	11,587	-	11,587	6,762	-	6,762
Chapter Fees	11,300	-	11,300	10,075	-	10,075
Ad Income	1,000	-	1,000	-	-	-
Administrative Fees	834	-	834	759	-	759
Mailing List Rental	723	-	723	1,013	-	1,013
Publications	347	-	347	909	-	909
Investment Income	-	1,483	1,483	-	(766)	(766)
Interest Income	345	160	505	357	382	739
Net Assets Released from Restrictions	4,205	(4,205)	-	4,788	(4,788)	-
Total Support and Revenues	373,773	3,860	377,633	295,963	125	296,088
Expenses						
Program Services						
General	187,521	-	187,521	208,677	-	274,524
Meetings	78,968	-	78,968	52,079	-	51,751
Publications	41,541	-	41,541	44,191	-	51,595
Supporting Services						
Management and General	80,714	-	80,714	62,946	-	67,068
Membership Development	10,691	-	10,691	12,861	-	19,925
Total Expenses	399,435	-	399,435	380,754	-	464,863
Change in Net Assets	(25,662)	3,860	(21,802)	(84,791)	125	(84,666)
Net Assets, Beginning of Year	28,906	121,244	150,150	113,697	121,119	234,816
Net Assets, End of Year	$ 3,244	$ 125,104	$ 128,348	$ 28,906	$ 121,244	$ 150,150

See Accompanying Notes.

LABOR AND EMPLOYMENT RELATIONS ASSOCIATION
STATEMENTS OF CASH FLOWS
FOR THE YEARS ENDED DECEMBER 31, 2014 AND 2013

	2014	2013
CASH FLOWS FROM OPERATING ACTIVITIES		
Change in Net Assets	$ (21,802)	$ (84,666)
Adjustments to Reconcile Change in Net Assets to Net Cash Provided by (Used in) Operating Activities		
Depreciation Expense	1,080	1,273
Loss on Disposal of Fixed Assets	445	-
Net Unrealized (Gains) Losses on Investments	(715)	1,639
(Increase) Decrease in Operating Assets		
Accounts Receivable, Net	26,306	(1,070)
Prepaid Expenses	481	(290)
Inventory	1,060	(182)
Accrued Interest and Royalties	(659)	(3,375)
Increase (Decrease) in Operating Liabilities		
Accounts Payable	(8,004)	242
Funds Held for Other Organizations	2,581	380
Accrued Liabilities	2,256	(7,934)
Dues Collected in Advance	(22,319)	4,563
Subscriptions Collected in Advance	(3,018)	(542)
Deferred Sponsorships	25,000	-
Deferred Chapter Dues	860	2,500
Other Deferred Revenue	855	900
Net Cash Provided by (Used in) Operating Activities	4,407	(86,562)
CASH FLOWS FROM INVESTMENT ACTIVITIES		
Reinvested Interest, Dividends, and Capital Gains	(768)	(873)
Payments for Property and Equipment	-	(539)
Net Cash Provided by (Used in) Investment Activities	(768)	(1,412)
CASH FLOWS FROM FINANCING ACTIVITIES		
None	-	-
Net Cash Provided by (Used in) Financing Activities	-	-
Net Increase (Decrease) in Cash and Cash Equivalents	3,639	(87,974)
Beginning Cash and Cash Equivalents	223,184	311,158
Ending Cash and Cash Equivalents	$ 226,823	$ 223,184

See Accompanying Notes.

LABOR AND EMPLOYMENT RELATIONS ASSOCIATION
NOTES TO FINANCIAL STATEMENTS
DECEMBER 31, 2014 AND 2013

NOTE 1 – NATURE OF ACTIVITIES AND SIGNIFICANT ACCOUNTING POLICIES

Organization and Nature of Activities

The Labor and Employment Relations Association (LERA) was founded in 1947 to encourage research in all aspects of the field of labor, employment, and the workplace. It is a nonprofit scholarly association of academic, labor, business and neutral communities committed to the full discussion and exchange of ideas between and among its broad constituencies through meetings, publications, and its various electronic listservs and websites. The LERA National Office is located in Champaign, Illinois and serves the association by planning conferences and meetings and publishing the various researches of its members. The main sources of support and revenues for LERA are contributions, membership dues and subscriptions, and meeting income.

Subsequent Events

Subsequent events have been evaluated through May 19, 2015, which is the date the financial statements were available to be issued.

Accrual Basis of Reporting

LERA has chosen to report on the accrual basis of accounting. Accordingly, revenue is recognized when earned and expenses are recognized when incurred in conformity with accounting principles generally accepted in the United States of America (GAAP).

Estimates

The preparation of financial statements in conformity with GAAP requires management to make estimates and assumptions that affect certain reported amounts and disclosures. Accordingly, actual results could differ from those estimates.

Cash and Cash Equivalents

For purposes of reporting cash flows, cash and cash equivalents include all cash and highly liquid investments acquired with an original maturity date of three months or less. Since the penalties of converting certificate of deposits to cash is insignificant, all certificate of deposits have been included with cash and cash equivalents. As of December 31, 2014 and 2013, $102,248 and $99,871 of cash and cash equivalents are temporarily restricted and are subject to donor-imposed stipulations.

Investments

Investments consist of intermediate term bond index funds. All investments are held as temporarily restricted and are subject to donor-imposed stipulations.

Accounts Receivable

Accounts receivable are recorded primarily for outstanding invoices for membership dues and subscriptions. An allowance for doubtful accounts is based on an analysis of expected collection rates determined from experience. LERA had an allowance for doubtful accounts of $100 at December 31, 2014 and 2013.

Inventory

LERA's inventory of directories, research volumes, proceedings, and perspective magazines is carried at the lower of cost and market value. Cost is determined on the basis of first in – first out.

Property and Equipment

Property and equipment expenditures in excess of $500 are capitalized at cost. Donated property and equipment are capitalized at estimated cost or fair market value at the time of donation. Depreciation of the assets is computed using the straight-line method over their estimated useful lives. The range of estimated useful lives by type of asset is as follows:

Furniture and Equipment	5-7 years

Net Assets

Net assets of LERA and changes therein are classified and reported as follows:

Unrestricted Net Assets – Net assets that are not subject to donor-imposed stipulations.

Temporarily Restricted Net Assets – Net assets subject to donor-imposed stipulations that may or will be met, either by actions of LERA and/or passage of time. When a restriction expires, temporarily restricted net assets are reclassified from restrictions. Contributions received with temporarily restrictions that are met in the same reporting period are reported as unrestricted support and increase unrestricted net assets. LERA had $125,104 and $121,244 of temporarily restricted net assets as of December 31, 2014 and 2013, respectively.

Permanently Restricted Net Assets – Net assets subject to donor-imposed stipulations that they be maintained permanently by LERA. Generally, the donors of these assets permit LERA to use all or part of the income earned on any related investments for general or specific purposes. LERA had no permanently restricted net assets as of December 31, 2014 and 2013.

Grant Revenue

Grant funds received under contracts that require provision of services by LERA are recognized as revenue to the extent that expenses have been incurred for the purpose specified by the grantor. The excess of grant monies received over expenses incurred is accounted for as deferred revenue. The excess of grant expenditures over grant monies received is accounted for as grant revenue receivable to the extent that additional grant revenue has been awarded and will be paid.

Grant funds received with only general restrictions and public donations are considered to be available for unrestricted use unless specifically restricted by the donor. When such restrictions exist, the revenue is recorded as temporarily restricted until the restrictions are met.

Contributed Services and Goods

Contributed services are reported as contribution revenue and as assets or expenses only if the services create or enhance a non-financial asset (for example property and equipment) or:

- Would typically need to be purchased by LERA if the services had not been provided by contribution.
- Require specialized skills.
- Are provided by individuals with those skills (such as accounting, financial, construction, educational, electrical, legal, medical, and other services provided by accountants, investments advisers, contractors, teachers, electricians, lawyers, doctors, and other professional and craftspeople).

For the years ended December 31, 2014 and 2013, the value of contributed services meeting the defined requirements for recognition in the financial statements as outlined above were not material and have not been recorded on the financial statements. LERA does receive free office space from the University of Illinois. However, the value for the use of this office space has not been reported on the financial statements.

Income Tax Status

LERA is a nonprofit association that is exempt from federal income tax under Section 501 (c)(3) of the Internal Revenue Code. In addition, the Internal Revenue Service has determined that LERA is not a private foundation as defined in Section 509(a)(1) and Section 170(b)(1)(A)(vi) of the Code.

LERA has evaluated its exposure resulting from uncertain income tax position and determined the exposure is not material to the financial statements. In addition, LERA is not aware of any tax position for which a significant change is reasonably possible within the next 12 months. Therefore, these financial statements do not include a liability for uncertain tax positions. Upon recognition of a liability for an uncertain tax position, LERA would recognize interest expense and penalties in operating expenses.

LERA files information tax returns in the U.S. federal jurisdiction and the state of Illinois. Its federal and Illinois information tax returns prior to fiscal year 2011 are closed. LERA does not have any tax returns currently under examination by either the Internal Revenue Service (IRS) or any U.S. state jurisdiction.

Membership Dues and Advance Subscriptions Collected

Membership dues and subscriptions are assessed and recognized as revenue based on the life of the dues or subscription.

Functional Allocation of Expenses

The costs of providing the various programs and other activities have been summarized on a functional basis in the statement of activities. Accordingly, certain costs have been allocated among the programs and supporting services benefited.

NOTE 2 – ARRANGEMENTS WITH THE UNIVERSITY OF ILLINOIS

LERA moved its offices to the University of Illinois at the end of 1999. Under an arrangement with the University, the employees of LERA are employed by the University. The employees' pension and benefits are part of the University's plans. LERA then reimburses the University quarterly for the cost of its employees.

The University of Illinois holds some cash for LERA. These "claim on cash" balances were $21,614 and $29,804 as of December 31, 2014 and 2013, respectively. These balances have been included with cash and cash equivalents.

NOTE 3 – UCIRHRP SERVICING ARRANGEMENT TO ADMINISTER BUSINESS

On January 3, 2003, LERA entered into a servicing arrangement to administer the business of the University Council of Industrial Relations and Human Resources Programs (UCIRHRP). These servicing duties include dues notification and collection, annual meeting arrangement and report preparation, and maintaining a data base and network communications for this separate organization. A separate financial statement is maintained for this company, but UCIRHRP's cash is maintained in LERA's general bank account. The amount of cash belonging to UCIRHRP in the general account is shown on LERA's balance sheet as Funds Held for Other Organizations. UCIRHRP's cash balance in LERA's custody was $5,865 at December 31, 2014 and $4,935 at December 31, 2013. LERA is allowed to keep 20% of dues collected each year as an administrative fee.

NOTE 4 – CONCENTRATIONS OF CREDIT RISK

Financial instruments that potentially subject LERA to credit risk consist principally of checking accounts, money markets accounts, and certificates of deposits at financial institutions. However, management continuously monitors LERA's balances at financial institutions. Deposits held at any financial institutions were fully insured by the Federal Deposit Insurance Corporation (FDIC) at December 31, 2014 or 2013. The "claims on cash" held with the University of Illinois as stated in Note 2 is not insured since it is not held in a financial institution. In addition, LERA had undeposited funds of $7,850 and $4,665 at December 31, 2014 and 2013, respectively. Since these funds were not yet deposited with a financial institution it was covered by FDIC.

NOTE 5 – TEMPORARILY RESTRICTED NET ASSETS

Susan C. Eaton Scholar-Practitioner Memorial Fund

LERA set up a memorial fund in honor of an author of a 1998 "Perspectives on Work" article and active young LERA member, who died on December 30, 2003. At its June 1, 2004 meeting, LERA approved the establishment of an annual Susan C. Eaton Scholar-Practitioner award and grant to be paid to one or more qualified scholar researchers in even-numbered years or practitioners in odd-numbered years doing research in the labor and employment relations or related field. LERA's Executive Board directed that 10% of any gifts received each year are unrestricted and can be used for administrative expenses. As of December 31, 2014 and 2013, these temporary restricted funds totaled $31,935 and $34,354, respectively.

Kochan-Sleigh Best Dissertation Award Fund

On March 27, 2006, LERA set up a temporarily restricted fund with contributions from a member and matching contributions from General Electric. The fund is restricted for a minimum of fifteen years and may be used to pay for the $1,000 annual best dissertation award and plaque. After the fifteen year period, the fund may be continued or the amount remaining in the fund may be released and become available to LERA for unrestricted purposes. LERA's Executive Board directed that 10% of any gifts received each year are unrestricted and can be used for administrative expenses. As of December 31, 2014 and 2013, these temporary restricted funds totaled $43,624 and $44,765, respectively.

Jim Scoville Award Fund

On January 9, 2009, LERA set up a temporarily restricted fund with a $10,000 contribution from the University of Minnesota's Industrial Relations Center to honor a member and retiring professor. The fund was established to pay an annual award of $500 and a plaque for best paper on international and comparative employment issues. LERA's Executive Board directed that 10% of any gifts received each year are unrestricted and can be used for administrative expenses. As of December 31, 2014 and 2013, these temporarily restricted funds totaled $7,041 and $7,031, respectively.

John T. Dunlop Public Policy Fund

On May 8, 2010, LERA established a temporarily restricted fund to subsidize a named John T. Dunlop Public Policy Session at future LERA National Policy Forums and/or at LERA annual meetings. LERA's Executive Board directed that 10% of any gifts received each year are unrestricted and can be used for administrative expenses. This fund will be so named for five or ten years, at which time LERA's Executive Board will review the fund to determine whether to continue or modify the fund. As of December 31, 2014 and 2013, these temporary restricted funds totaled $19,474 and $17,331, respectively.

Gladys and Walter Gershenfeld Publication Fund

On May 8, 2010, LERA established a temporarily restricted fund to honor the memory of long-time LERA members and LERA Past President Walter and Gladys Gershenfeld for the purpose of supporting LERA's electronic and print publications. The fund will be so named for ten years, at which time LERA will review the fund's purpose and uses, and whether to continue or modify the fund. LERA's Executive Board directed that 10% of any gifts received each year are unrestricted and can be used for administrative expenses. As of December 31, 2014 and 2013, these temporarily restricted funds totaled $23,030 and $17,763, respectively.

NOTE 6 – INVESTMENTS

Financial Accounting Standards Board ("FASB") Codification 820, Fair Value Measurements, establishes a framework for measuring fair value. That framework provides a fair value hierarchy that prioritizes the inputs to valuation techniques used to measure fair value. The hierarch gives the highest priority to unadjusted quoted prices in active markets for identical assets or liabilities (Level 1 measurements) and the lowest priority to unobservable inputs (Level 3 measurements). The three levels of the fair value hierarchy under FASB Codification 820 are described below:

> Level 1 – Inputs to the valuation methodology are based on unadjusted quoted prices for identical assets or liabilities in active markets that LERA has the ability to access.
>
> Level 2 – Inputs to the valuation methodology include quoted prices for similar assets or liabilities in active markets, quoted prices for identical or similar assets or liabilities in inactive markets, inputs other than quoted prices that are observable for the asset or liability, and inputs that are derived principally from or corroborated by observable market data by correlation or other means. If the asset or liability has a specified (contractual) term, the Level 2 input must be observable for substantially the full term of the asset or liability.
>
> Level 3 – Inputs to the valuation methodology are unobservable and significant to the fair value measurement.

The asset's and liability's fair value measurement level within the fair value hierarchy is based on the lowest level of any input that is significant to the fair value measurement. Valuation techniques used need to maximize the use of observable inputs and minimize the use of unobservable inputs.

The following is a description of the valuation methodologies used for assets measured at fair value. There have been no changes in the methodologies used at December 31, 2014 and 2013.

> *Intermediate Term Bond Funds:* Valued at the NAV of shares held by the Organization at year end based on readily determinable fair values, which are published daily and are the basis for current transactions.

The preceding methods described may produce a fair value calculation that may not be indicative of net realizable value or reflective of future fair values. Furthermore, although LERA believes its valuation methods are appropriate and consistent with other market participants, the use of different methodologies or assumptions to determine the fair value of certain financial instruments could result in different fair value measurements at the reporting date.

The following table sets forth by level, within the fair value hierarchy, LERA's assets at fair value as of December 31, 2014 and 2013:

Assets at Fair Value as of December 31, 2014

	Level 1	Level 2	Level 3	Total
Intermediate Term Bond Funds	$22,856	-	-	$22,856

Assets at Fair Value as of December 31, 2013

	Level 1	Level 2	Level 3	Total
Intermediate Term Bond Funds	$21,373	-	-	$21,373

Net investment earnings for the years ended December 31, 2014 and 2013 are summarized as follows:

	2014	2013
Dividends	$ 636	$ 637
Capital Gains Distributions	132	236
Net Unrealized Gains (Losses)	715	(1,639)
Total Investment Income	$ 1,483	$ (766)

This investment income, including unrealized gains and losses, are being reported on the statement of activity as investment income.

NOTE 7 – PROPERTY AND EQUIPMENT

As of December 31, 2014, Property and Equipment consists of:

Furniture and Equipment	$ 15,639
Less: Accumulated Depreciation	(14,052)
Property and Equipment, Net	$ 1,587

As of December 31, 2013, Property and Equipment consists of:

Furniture and Equipment	$ 17,347
Less: Accumulated Depreciation	(14,235)
Property and Equipment, Net	$ 3,112

Depreciation expense for the years ended December 31, 2014 and 2013 was $1,080 and $1,273, respectively.

NOTE 8 – ACCRUED EXPENSES

As of December 31, 2014, Accrued Expenses consist of:

Accrued Payroll	$ 5,082
Accrued Benefits/Taxes	2,002
Accrued Vacation	17,231
Total	$ 24,315

As of December 31, 2013, Accrued Expenses consist of:

Accrued Payroll	$ 5,879
Accrued Benefits/Taxes	2,118
Accrued Vacation	14,062
Total	$ 22,059

NOTE 9 – COMMITMENTS

LERA had a contract with University of Illinois Press (UIP) for UIP to provide membership management services and publication of *Perspectives on Work* (magazine) for an annual fee of $28,280. The five year agreement commenced on January 1, 2007 and expired on December 31, 2012. Although the original contract expired, the terms continued through December 31, 2014. As of December 31, 2014, LERA is no longer under contract with UIP and no future payments are due to UIP.

On February 26, 2013, LERA signed a 12 month contract from March 1, 2013 to February 28, 2014 with a company to provide services to manage the EPRN website and provide content for the LERA website. The contract states that LERA will pay this company $2,400 per month for a total of $28,800. The Agreement automatically renews for consecutive 12 month terms, unless the Agreement is terminated by one or both of the parties as set forth herein. Beginning the second cycle and continuing until termination or renegotiation, LERA will increase the base rate every year by two percent (2%) over the previous year's base rate. For example: $2,448 per month for twelve months starting March 1, 2014; $2,497 per month for twelve months starting March 1, 2015; $2,546 per month for twelve months starting March 1, 2016; and so on.

LERA has signed contracts with hotels in Portland, Pittsburgh, and Minneapolis for the 2014, 2015, and 2016 Annual Meetings. If these contracts were cancelled at December 31, 2014 and 2013, LERA would have owed $61,965 and $76,903, respectively, to the Portland, Pittsburgh, and Minneapolis hotels.

NOTE 10 – PENSION PLAN AND RETIREMENT CONTRIBUTIONS

The University of Illinois (the University) contributes to the State Universities Retirement System of Illinois (SURS), a cost-sharing multiple-employer, defined-benefit pension plan with a special funding situation whereby the State makes substantially all actuarially determined required contributions on behalf of the participating employers. SURS was established July 21, 1941, to provide retirement annuities and other benefits for staff members and employers of the State universities, certain affiliated organization, and certain other State educational and scientific agencies and for survivors, dependents and other beneficiaries of such employees. SURS is considered a component unit of the State of Illinois' financial reporting entity, and is included in the State's financial reports as a pension trust fund. SURS is governed by Section 5/15, Chapter 40, of the *Illinois Compiled Statutes*. SURS issues a publically available financial report and required supplementary information. That report may be obtained by accessing the Web site at www.SURS.org or by calling 1-800-275-7877.

SURS federal tax identification number is 37-6006008. The plan was underfunded by $20,110,500,000 as of June 30, 2013, the most recent actuarial valuation date, which includes actuarial value of assets totaling $14,262,600,000 and accrued actuarial liabilities totaling $34,373,100,000. The plan was underfunded by $21,584,800,000 as of June 30, 2014, the most recent actuarial valuation date, which includes actuarial value of assets totaling $12,788,300,000 and accrued actuarial liabilities totaling $34,373,100,000.

While LERA reports under Financial Accounting Standards Board, the University reports under Governmental Accounting Standards Board (GASB). GASB Statement No. 68, *Accounting and Financial Reporting for Pensions*, issued June 2012, will be effective for the University beginning with its year ending June 30, 2015. This statement requires governments providing defined benefit pensions to recognize their long-term obligation for pension benefits as a liability for the first time, and to more comprehensively and comparably measure the annual costs of pension benefits. This Statement also enhances accountability and transparency through revised and new note disclosures and required supplementary information. As LERA participates in SURS through the University, this standard may impact LERA's financial statements in addition to the University. Management is in the process of evaluating the potential impact this standard will have on its financial statements.

Eligible employees must participate upon initial employment. Employees are ineligible to participate if (a) employed after having attained age 68; (b) employed less than 50% of full time; or (c) employed less than full time and attending classes with an employer.

Plan members are required to contribute 8.0% of their annual covered salary and substantially all employer contributions are made by the State on behalf of the individual employers at an actuarially determined rate. LERA's current employer rate is 11.71% of annual covered payroll. The contribution requirements of plan members and employers are established and may be amended by the Illinois General Assembly. During the years ended December 31, 2014, 2013 and 2012, LERA reimbursed the University of Illinois $14,743, $16,436, and $20,632, respectively, for the employer's payments to SURS for eligible employees.

Employers hired prior to April 1, 1986 are exempt from contributions required under the Federal Insurance Contribution Act. Employees hired after March 31, 1986 are required to contribute 1.45% of their gross salary for Medicare. LERA is required to match this contribution. As of December 31, 2014 and 2013, all employees were required to contribute 1.45% of their gross salary for Medicare.

Employees may also elect to participate in certain tax-sheltered retirement plans. These voluntary plans permit employees to designate a part of their earnings into tax-sheltered investments and thus defer federal and state income taxes on their contributions and accumulated earnings under the plans. Participation and the level of employee contributions are voluntary. The employer is not required to make contributions to these plans.

NOTE 11 – POST-EMPLOYMENT BENEFITS

The State of Illinois (the State) provides health, dental, vision, and life insurance benefits for retirees and their dependents in a program administered by the Department of Central Management Services. Substantially all State employees become eligible for post-employment benefits if they eventually become annuitants of one of the State sponsored pension plans. Health, dental, and vision benefits include basic benefits for annuitants and dependents under the State's self-insurance plan and insurance contracts currently in force. Annuitants may be required to contribute toward health, dental, and vision benefits with the amount based on factors such as date of retirement, years of credited service with the State, whether the annuitant is covered by Medicare, and whether the annuitant has chosen a managed health care plan. Annuitants who retired prior to January 1, 1998, and who are vested in the SURS do not contribute toward health, dental, and vision benefits. For annuitants who retired on or after January 1, 1998, the annuitant's contribution amount is reduced five percent for each year of credited service with the State; therefore, those annuitants with twenty or more years of credited service do not have to contribute toward health, dental, and vision benefits. Annuitants also receive life insurance coverage equal to the annual salary of the last day of employment until age 60, at which time the benefit becomes $5,000.

The State pays the SURS' portion of employer costs for the benefits provided. The total cost of the State's portion of health, dental, vision, and life insurance benefits of all members, including postemployment health, dental, vision, and life insurance benefits, is recognized as an expenditure by the State in the Illinois Comprehensive Annual Financial Report. The State finances the costs on a pay-as-you-go basis. The total costs incurred for health, dental, vision, and life insurance benefits are not separated by department or component unit for annuitants and their dependents nor active employees and their dependents.

A summary of post-employment benefit provisions, changes in benefit provisions, employee eligibility requirements including eligibility for vesting, and the authority under which benefit provisions are established are included as an integral part of the financial statements of the Department of Central Management Services. A copy of the financial statements of the Department of Central Management Services may be obtained by writing to the Department of Central Management Services, Stratton Building, Room 720, Springfield, Illinois, 62706-4100.

LABOR AND EMPLOYMENT RELATIONS ASSOCIATION
SCHEDULE OF FUNCTIONAL EXPENSES
FOR THE YEAR ENDED DECEMBER 31, 2014

		MEETINGS			PUBLICATIONS						SUPPORTING SERVICES		
	General	Annual Conference	ASSA Meeting	Other Meetings	Directory	Newsletter	Annual Proceedings	Perspectives	Research Volume	Other Publications	Management & General	Membership Development	Totals
Compensation	$ 112,163	$ -	$ -	$ -	$ -	$ -	$ -	$ -	$ -	$ -	$ 16,173	$ -	$ 128,336
Payroll Taxes & Fringes	45,087	-	-	-	-	-	-	-	-	-	6,706	-	51,793
Contract Labor	29,280	-	-	-	-	-	-	-	-	-	-	-	29,280
Depreciation	-	-	-	-	-	-	-	-	-	-	1,080	-	1,080
Loss on Disposal of Fixed Assets	-	-	-	-	-	-	-	-	-	-	445	-	445
Insurance	-	-	-	-	-	-	-	-	-	-	3,773	-	3,773
Bank Charges	-	-	-	-	-	-	-	-	-	-	2,731	-	2,731
Promotion	-	4,256	139	-	-	-	-	-	-	-	-	3,252	7,647
Fulfillment	-	-	-	-	-	-	-	-	-	-	20,240	-	20,240
Postage and Freight	-	707	52	-	-	-	-	1,403	1,110	-	228	-	3,500
Accounting/Auditing	-	-	-	-	-	-	-	-	-	-	4,800	-	4,800
Printing and Production	-	1,841	-	-	-	-	-	12,732	4,400	68	-	-	19,041
Services	-	-	-	-	-	-	487	12,737	6,715	-	-	-	19,939
Other Publication Costs	-	-	-	-	875	375	500	139	-	-	-	-	1,889
Meals/Receptions	-	47,666	-	-	-	-	-	-	-	-	-	-	47,666
Travel	-	3,602	1,889	611	-	-	-	-	-	-	-	-	6,102
Other Meeting Expenses	-	15,171	479	150	-	-	-	-	-	-	-	-	15,800
Computer Supplies/Services	-	2,405	-	-	-	-	-	-	-	-	439	-	2,844
Website	-	-	-	-	-	-	-	-	-	-	18,601	-	18,601
Office Supplies	991	-	-	-	-	-	-	-	-	-	1,263	-	2,254
Student & Member Awards	-	-	-	-	-	-	-	-	-	-	-	5,985	5,985
Fundraising Expense	-	-	-	-	-	-	-	-	-	-	-	994	994
Duplicating Expense	-	-	-	-	-	-	-	-	-	-	367	-	367
Telephone & Fax	-	-	-	-	-	-	-	-	-	-	438	-	438
Chapter Expenses	-	-	-	-	-	-	-	-	-	-	-	349	349
Dues	-	-	-	-	-	-	-	-	-	-	641	-	641
Other Committee Expenses	-	-	-	-	-	-	-	-	-	-	-	111	111
Miscellaneous Office	-	-	-	-	-	-	-	-	-	-	2,789	-	2,789
	$ 187,521	$ 75,648	$ 2,559	$ 761	$ 875	$ 375	$ 987	$ 27,011	$ 12,225	$ 68	$ 80,714	$ 10,691	$ 399,435

LABOR AND EMPLOYMENT RELATIONS ASSOCIATION
SCHEDULE OF FUNCTIONAL EXPENSES
FOR THE YEAR ENDED DECEMBER 31, 2013

		MEETINGS			PUBLICATIONS						SUPPORTING SERVICES		
	General	Annual Conference	ASSA Meeting	Other Meetings	Directory	Newsletter	Annual Proceedings	Perspectives	Research Volume	Other Publications	Management & General	Membership Development	Totals
Compensation	$ 131,006	$ -	$ -	$ -	$ -	$ -	$ -	$ -	$ -	$ -	$ 14,396	$ -	$ 145,402
Payroll Taxes & Fringes	52,911	-	-	-	-	-	-	-	-	-	6,434	-	59,345
Contract Labor	24,000	-	-	-	-	-	-	-	-	-	-	-	24,000
Depreciation	-	-	-	-	-	-	-	-	-	-	1,273	-	1,273
Loss on Disposal of Fixed Assets	-	-	-	-	-	-	-	-	-	-	-	-	-
Insurance	-	-	-	-	-	-	-	-	-	-	3,653	-	3,653
Bank Charges	-	-	-	-	-	-	-	-	-	-	1,882	-	1,882
Promotion	-	8,111	-	-	-	-	-	-	-	-	-	4,395	12,506
Fulfillment	-	-	-	-	-	-	-	-	-	-	13,180	-	13,180
Postage and Freight	-	-	-	-	-	-	-	1,102	2,221	23	281	-	3,627
Accounting/Auditing	-	-	-	-	-	-	-	-	-	-	4,800	-	4,800
Printing and Production	-	1,238	-	-	-	-	-	17,893	5,415	60	-	-	24,606
Services	-	-	-	-	-	-	1,188	4,058	8,360	-	-	-	13,606
Other Publication Costs	-	-	-	-	1,750	750	1,000	352	-	19	-	-	3,871
Meals/Receptions	-	28,967	-	-	-	-	-	-	-	-	-	-	28,967
Travel	-	6,469	503	725	-	-	-	-	-	-	-	-	7,697
Other Meeting Expenses	-	4,687	114	-	-	-	-	-	-	-	-	-	4,801
Computer Supplies/Services	-	1,265	-	-	-	-	-	-	-	-	5,296	-	6,561
Website	-	-	-	-	-	-	-	-	-	-	9,432	-	9,432
Office Supplies	760	-	-	-	-	-	-	-	-	-	612	-	1,372
Student & Member Awards	-	-	-	-	-	-	-	-	-	-	-	6,097	6,097
Fundraising Expense	-	-	-	-	-	-	-	-	-	-	-	1,650	1,650
Duplicating Expense	-	-	-	-	-	-	-	-	-	-	1,047	-	1,047
Telephone & Fax	-	-	-	-	-	-	-	-	-	-	476	-	476
Chapter Expenses	-	-	-	-	-	-	-	-	-	-	-	292	292
Dues	-	-	-	-	-	-	-	-	-	-	65	-	65
Other Committee Expenses	-	-	-	-	-	-	-	-	-	-	-	427	427
Miscellaneous Office	-	-	-	-	-	-	-	-	-	-	119	-	119
	$ 208,677	$ 50,737	$ 617	$ 725	$ 1,750	$ 750	$ 2,188	$ 23,405	$ 15,996	$ 102	$ 62,946	$ 12,861	$ 380,754

The LERA gratefully acknowledges the continuing support of its Sustaining Sponsors and Organizational Members.

SUSTAINING SPONSORS 2014–2015

UAW/Ford National Programs
Bloomberg BNA
Institute for Construction Economic Research
Kaiser Permanente
Prouty Castrey Art Fund

ORGANIZATIONAL MEMBERS 2015

AFL-CIO Department of Professional Employees
AFL-CIO Working for America Institute
American Federation of Teachers
BlueCross BlueShield National Labor Office
Boston University, Questrom School of Business
Cornell University Scheinman Institute of Conflict Resolution
Cornell University School of Industrial and Labor Relations
Federal Mediation and Conciliation Service
Harvard University
Indiana University of Pennsylvania PA Center for the Study of Labor Relations
International Association of Machinists and Aerospace Workers
International Brotherhood of Boilermakers
Jobs with Justice Education Fund
Massachusetts Institute of Technology (MIT) Sloan School of Management
Michigan State University School of Labor and Industrial Relations
National Public Employer Labor Relations Association
New York State Nurses Association Economic & General Welfare Program
Orange County Transportation Authority
Pennsylvania State University Labor Studies and Employment Relations
Rollins College Department of Business & Social Entrepreneurship
Rutgers University School of Management and Labor Relations
Saint Joseph's University Haub School of Business
San Diego Municipal Employee Association
Seabury Group LLC
Society for Human Resource Management (SHRM)
Tennessee State Employees Association
The Dannon Company
UCLA Institute on Conflict Resolution
United Food and Commercial Workers Local #1776
United Steelworkers
University of California-Los Angeles, Institute for Research on Labor and Employment
University of California-Los Angeles, Anderson School of Management
University of California—Berkeley, Institute for Research on Labor and Employment
University of Illinois at Urbana-Champaign School of Labor and Employment Relations
University of Minnesota Carlson School of Management
University of Newcastle
University of Toronto Center for Industrial Relations and Human Resources
West Virginia University Department of Industrial Relations and Management

For more information about Organizational Membership with the LERA, please visit the LERA website: http://www.lera.illinois.edu/about/OrgMemInfo.html.

Alphabetical List of Authors

You are invited to become a member of

THE LABOR AND EMPLOYMENT RELATIONS ASSOCIATION

The Labor and Employment Relations Association (LERA) was founded in 1947 by a group who felt that the growing field of industrial relations required an association in which professionally minded people from different organizations could meet. It was intended to enable all who were professionally interested in industrial relations to become better acquainted and to keep up to date with the practices and ideas at work in the field. To our knowledge there is no other organization that affords the multiparty exchange of ideas we have experienced over the years–a unique and valuable forum. After 69 years, both our academic and practitioners agree with the conviction of the founders that the encouragement, reporting, and critical discussion of research are essential if our professional field is to advance.

Our local and national membership of 5,000 includes representatives of management, unions, government; practitioners in consulting, arbitration, mediation, and law; and scholars and teachers representing many disciplines in colleges and universities in the United States and Canada, as well as abroad. Libraries and institutions interested in the publications of the Association are also invited to become subscribing members and thereby receive the same member publications. Organizational memberships in the Association are also available.

Membership dues cover publications for 12 months and entitle members to the electronic newsletter, membership directory and proceedings of the annual meeting online, and printed copies of the annual research volume, and the magazine, *Perspectives on Work* (volumes also available online). Additional online resources include Chapter Profiles and an IR/HR Degree Program Listing for the United States, Canada, and Australia.

Dues for the 2015 calendar year are listed below. International memberships require an additional $20.00 per year to cover postage costs.

Regular Member	$195.00
Contributing Member	$300.00
Emeritus Member (10 years+ membership and no longer employed)	$105.00
Student Member (full-time, limited to 4 consecutive years)	$25.00
Library Subscriber	$250.00

Tax-deductible financial contributions to the Association to support its educational activities are always welcome. You can make an online contribution at our website, leraweb.org, and choosing "Donate". If you are not already a member, we invite you to join the LERA by sending your membership application and dues payment. You can also join online at the LERA website, leraweb.org by choosing "Membership", or request an invoice. Inquiries regarding membership, meetings, and publications should be addressed to the LERA office.

LABOR AND EMPLOYMENT RELATIONS ASSOCIATION
University of Illinois at Urbana-Champaign
121 LER Building, 504 E. Armory Avenue
Champaign, IL 61820 USA
Phone: 217-333-0072 Fax: 217-265-5130
Email: LERAoffice@illinois.edu Web: leraweb.org

Also in the
Labor and Employment Relations Association Series

Inequality, Uncertainty, and Opportunity: The Varied and Growing Role of Finance in Labor Relations
Edited by Christian E. Weller...$29.95

The Disunited States of America: Employment Relations Systems in Conflict
Edited by David Jacobs and Peggy Kahn...$29.95

Collective Bargaining Under Duress: Case Studies of Major U.S. Industries
Edited by Howard Stranger, Ann C. Frost, and Paul F. Clark...$29.95

Public Jobs and Political Agendas: The Public Sector in an Era of Economic Stress
Edited by Daniel J. B. Mitchell...$29.95

Employee Ownership and Shared Capitalism: New Direction in Research
Edited by Edward J. Carberry...$29.95

Transforming the U.S. Workforce Development System: Lessons from Research and Practice
Edited by David Finegold, Mary Gatta, Hal Salzman, and Susan J. Schurman.......................$29.95

Human Rights in Labor and Employment Relations: International and Domestic Perspectives
Edited by James A. Gross and Lance Compa...$29.95

The Gloves-off Economy: Workplace Standards at the Bottom of America's Labor Market
Edited by Annette Bernhardt, Heather Boushey, Laura Dresser, and Chris Tilly.......................$29.95

Employee Pensions: Policies, Problems, and Possibilities
Edited by Teresa Ghilarducci and Christian E. Weller...$29.95

Contemporary Issues in Employee Relations
Edited by David Lewin...$29.95

Book orders and request for prices and order forms for the above LERA publications should be sent to:

ILR Press, Cornell Univ., P.O. Box 6525, Ithaca, NY 14851
Phone 607-277-2211, Fax 607-277-6292
Web www.cornellpress.cornell.edu
Additional volumes are available through ILR Press.

Information about ordering printed copies of the *Proceedings of the LERA Annual Meeting* can be found online at https://lera.memberclicks.net/publications. Orders for printed copies of *Perspectives on Work* and other LERA publications should be sent to the LERA Office:

LERA, 121 LER Building, 504 E. Armory Ave., MC-504, Champaign, IL 61820
Phone 217-333-0072, Fax 217-265-5130
Email LERAoffice@illinois.edu, Web leraweb.org